HC 79 .T4 M3557 2007
Martin, R. Lee.
Techonomics : the theory of
industrial evolution

COLORADO

MOUNTAIN COLLEGE

Alpine Campus
Library
Bristol Hall
P.O. Box 774688
Steamboat Springs
CO 80477

TECHONOMICS
The Theory of Industrial Evolution

Industrial Innovation Series
Adedeji B. Badiru
The University of Tennessee, Knoxville, Tennessee

Published Titles

Handbook of Industrial and Systems Engineering
   Adedeji B. Badiru

Forthcoming Titles

Techonomics: The Theory of Industrial Evolution
   H. Lee Martin

Computational Economic Analysis for Engineering and Industry
   Adedeji B. Badiru & Olufemi A. Omitaomu

Industrial Project Management: Concepts, Tools and Techniques
   Adedeji B. Badiru

TECHONOMICS
The Theory of Industrial Evolution

H. Lee Martin, Ph.D.

Taylor & Francis
Taylor & Francis Group
Boca Raton London New York

CRC is an imprint of the Taylor & Francis Group,
an informa business

CRC Press
Taylor & Francis Group
6000 Broken Sound Parkway NW, Suite 300
Boca Raton, FL 33487-2742

© 2007 by Taylor & Francis Group, LLC
CRC Press is an imprint of Taylor & Francis Group, an Informa business

No claim to original U.S. Government works
Printed in the United States of America on acid-free paper
10 9 8 7 6 5 4 3 2

International Standard Book Number-10: 0-8493-7067-1 (Hardcover)
International Standard Book Number-13: 978-0-8493-7067-0 (Hardcover)

This book contains information obtained from authentic and highly regarded sources. Reprinted material is quoted with permission, and sources are indicated. A wide variety of references are listed. Reasonable efforts have been made to publish reliable data and information, but the author and the publisher cannot assume responsibility for the validity of all materials or for the consequences of their use.

No part of this book may be reprinted, reproduced, transmitted, or utilized in any form by any electronic, mechanical, or other means, now known or hereafter invented, including photocopying, microfilming, and recording, or in any information storage or retrieval system, without written permission from the publishers.

For permission to photocopy or use material electronically from this work, please access www. copyright.com (http://www.copyright.com/) or contact the Copyright Clearance Center, Inc. (CCC) 222 Rosewood Drive, Danvers, MA 01923, 978-750-8400. CCC is a not-for-profit organization that provides licenses and registration for a variety of users. For organizations that have been granted a photocopy license by the CCC, a separate system of payment has been arranged.

Trademark Notice: Product or corporate names may be trademarks or registered trademarks, and are used only for identification and explanation without intent to infringe.

Library of Congress Cataloging-in-Publication Data

Martin, H. Lee.
   Techonomics : the theory of industrial evolution / H. Lee Martin.
     p. cm. -- (Indusrial innovation series ; 2)
   Includes bibliographical references and index.
   ISBN 0-8493-7067-1 (alk. paper)
   1. Technological innovations--Economic aspects. 2. Industrial organization (Economic theory) 3. Technological forecasting. 4. Evolutionary economics. 5. Social evolution. I. Title. II. Title: theory of industrial evolution.

HC79.T4M3557 2006
338'.064--dc22   2006045640

Visit the Taylor & Francis Web site at
http://www.taylorandfrancis.com

and the CRC Press Web site at
http://www.crcpress.com

Preface

March 17, 1986. A hospital room in the little town of Murfreesboro, Tennessee. My father had suffered nearly 12 years with the debilitating effects of emphysema. His strength was waning. Though I did not know at the time that this would be my last talk with him, we both sensed this was a special moment.

Dad had just celebrated his 75th birthday, and I had brought him a draft copy of my engineering dissertation on an obscure topic. He left college during the Great Depression to take a job — one of the few available at the time — and he had never returned to school. But I knew education was important to him, and I always felt my pursuit of a Ph.D. was as much to recoup his missed opportunity as it was my personal knowledge quest.

This was a special day. I saw in his eyes a father's pride in his son's accomplishment. He knew I was about to cross the doctoral finish line. It was one of those moments when the bond between father and son is at fullest strength, and my own heart filled with gratitude for the sacrifices of my father that made this day possible. All my life I had been amazed at his discipline and dedication to his work, but I had never asked him much about it. But on this day, at this time, I wanted to know.

You see, for the last 15 years of his career, the portion I was aware of, my father had been the area superintendent for electric power distribution in Middle Tennessee for the Tennessee Valley Authority. He was 24/7/365 long before the term was coined. Whenever there was a storm and the lights would go out, he would get a call and be off into the rain with his crews, finding and fixing the problem. His devotion to his job dominated his life, and I wanted to find out *why* before it was too late to ask.

"Dad, what made you work so hard in your career?"

At first I got the typical answer men of his generation seem to give: "Son, I wanted to provide well for your Mom and you two boys." I thanked him, but the answer was not satisfying, did not account for the deep dedication with which he worked. I pressed harder.

"But Dad, you were so passionate about keeping the lights on; seems to me it had to be something deeper."

I expected thoughtful hesitation, but he responded instantly telling me a story I have never forgotten. "Son, on a clear autumn afternoon about dusk, take U.S. Highway 231 north to Lebanon and sit on the ridge about 2 miles south of the city and you'll understand."

"Understand what, Dad?"

"Just sit there and look. In about 5 minutes, you'll see a light come on.... Then you'll understand."

"Help me, Dad. Understand what?"

"Just wait. In about 10 minutes you'll see another light, and then another and another. And you'll understand."

"Huh?"

"After about 5 more minutes, you'll see more lights than you can count, and you'll understand, Son. My team and I positively touched the lives of thousands of people we'll never meet — that made all the difference."

Those few moments changed my life. Within 3 months I left my secure position as a development engineer in the nation's largest federal lab for a downstairs office in my home, pursuing the challenge of a technology startup. I had realized that my ideas, my creations, my work product would never touch others as long as they were developed behind the massive fence of secrecy and bureaucracy surrounding most developments in a federal laboratory. After that conversation with my father, fear of the unknown business world was less daunting than knowing my career might be spent developing obscure devices that never touched another life.

In the twenty years that have come and gone since that March day, I have learned many lessons while striving to positively touch the lives of thousands of people I may never meet. TeleRobotics International, a company of two, began in 1986 in a residence in rural Tennessee. Following technology trends from robotics to imaging, hardware to software, mainframes to personal computers, and dedicated intranets to the World Wide Web, the company evolved with the demands and opportunities of the new economy. Ultimately, the company developed a 360° image that was utilized worldwide for interactive viewing of homes, hotels, automobiles, and other items of interest. Riding the massive technical and economic waves of the Internet boom, the company went public as iPIX in 1999, at one point reaching a valuation in excess of $2 billion. *Techonomics* is the essence of my anticipatory thought process gained during this journey. This book contains my hard-earned observations of the evolution of organizations during a period of unprecedented global technological and economic expansion.

Four authors' books have significantly shaped the conceptual development of *Techonomics:* Larry Downes and Chunka Mui, authors of *Unleashing the Killer App,* John Naisbitt, author of *Megatrends,* Jim Collins, author of *Good to Great,* and Charles Darwin, author of *The Origin of Species*.[1-4] I stand on the shoulder of these giants as I scan history attempting to bring understanding from the societal chaos instigated repeatedly by overwhelming technological innovation. It is my sincere desire to encourage your success by sharing and explaining many of the principles my techonomic journey has revealed.

REFERENCES

1. Downes, L. and Mui, C., *Unleashing the Killer App: Digital Strategies for Market Dominance,* revised ed., Harvard Business School Press, Boston, 1998.
2. Naisbitt, J., *Megatrends 2000,* reissue ed., Avon Books, New York, 1991.
3. Collins, J., *Good to Great: Why Some Companies Make the Leap ... and Others Don't,* HarperCollins, New York, 2001.
4. Darwin, C., On the Origin of Species by Means of Natural Selection, or the Preservation of Favoured Races in the Struggle for Life, Wordsworth Editions Limited, Hertfordshire, 1998.

Foreword

I first came to know Dr. Martin in 1998, when I was Commissioner of Economic Development for the State of Tennessee. Tennessee clearly faced a gap between our outstanding technology research resources, such as the Oak Ridge National Lab, and the commercial economic impact of marketable products and services that might result from these technologies. In short, we needed a specialized, targeted effort to bridge this gap.

With the creative thought of Alex Fischer, my Deputy Commissioner, and the support of then Governor Don Sundquist, we determined that an entity separate from state government could be more agile and aggressive in addressing this need in our state. We therefore created the Tennessee Technology Development Corporation (TTDC) for the purpose of moving technology from the research phase to the commercial phase for the economic benefit of our citizens.

We needed a special person to lead this effort, someone who had achieved this very result in his own professional life. As we looked across the state, it became clear that we needed to enlist the efforts of Dr. Lee Martin to lead the Tennessee Technology Development Corporation. Fortunately for us, he agreed. Lee brought with him the fundamental principle of understanding trends in technology and their potential economic effects. Exactly what we needed.

This effort was singularly successful, being guided by a diverse board of directors with expertise in technology, business, and government. The multi-year TTDC effort resulted in an inventory and assessment of our state research capabilities that could be brought to bear on the goal of producing job-creating and wealth-creating businesses. This was a huge step forward for Tennessee, and we are still reaping the benefits of Dr. Martin's efforts.

Today, as we witness the undeniable trends of globalization in our economy, it is clear that a techonomic approach is absolutely essential to our success. In order for our labor force, our companies, and our nation's economy to remain competitive in the new world economy, we simply must be the best at using technology to create economic value. There are other countries with cheaper labor, greater natural resources, or larger markets. However, to date, no country has greater technological assets or, more importantly, the demonstrated ability to convert those technological assets into products and services with economic value.

Nowhere is this more evident than in the energy sector, an area where I have spent most of my business and public service career. Our extensive dependence on foreign oil is an undeniable economic weakness and competitive disadvantage. Further, our substantial energy asset — perhaps 200 years of coal — will lose its economic value if we are not more successful with clean coal and clean air technologies.

I agree with Dr. Martin's prediction that nuclear energy is on the verge of rebirth in the U.S. — something that bodes well for our continued economic success. We have proven technologically that we can operate nuclear power plants very safely, producing low-cost electricity with no environmental emissions related to air quality, and we have further demonstrated that we can safely handle the waste product from this energy-producing process. We will soon see "next generation" nuclear plants that will be smaller, cheaper, and simpler to operate, producing perhaps only a fraction of the net waste product.

Great progress is also being made in the area of distributed energy resources, such as photovoltaics, passive solar, hybrid solar lighting, wind, and biomass. These technologies have their pros and cons, and continued technological progress is required, but they remain promising. And the whole concept of a hydrogen-based economy is just now beginning to be explored. Dr. Martin's description of energy as a fundamental component of every organism and organization, as well as an index of its health and success, is accurate and instructive. His explanation of its role in techonomic trends is fundamental to understanding the vital role of energy technologies and choices for our state and nation.

Finally, it is an important contribution of this book to so powerfully make the point that technological progress absolutely depends upon free markets. If we stifle competition in the marketplace and the rewards available in a capitalistic economy, technological progress will grind to a halt. America must make the case for this sine qua non of prosperity for all.

Reading this book and coming to understand the principles of techonomics will give you and your company or organization a competitive advantage. Anticipating technological trends and their economic impact will be the key to financial success for businesses and economic success for nations in the foreseeable future. Dr. Martin has done us a great service by articulating the principles of techonomics and the resulting analytical tools it provides. The wise reader will put them to use immediately.

Bill Baxter
Chairman of the Board
Tennessee Valley Authority

Abstract — Techonomics: Theory of Industrial Evolution

Have you ever wondered what are the forces behind globalization, mass customization, just-in-time delivery, virtual companies, and perfect information? Would you like to have the ability to spot and capitalize on economic trends before they reach the masses? Do you realize that, throughout history, military development of technology has been the leading indicator of future commercial applications? *Techonomics* explains the significant relationship between technology, economy, and organizations, providing you a worldview to understand and navigate our rapidly advancing world.

Techonomics provides a theory for organizational evolution that parallels Darwin's theory of organic evolution. Technological best practices tested by free-market competition determine successful organizations similar to the way favorable mutations in a competitive natural environment select the fittest organisms. This techonomic mindset gives the reader a platform to observe the rapid changes in our economy and provide insight into wise deployment of limited resources.

Successful entrepreneur and prolific inventor Dr. H. Lee Martin shares a technologist's marketplace insights gained from a 15-year journey from the garage to the public market. By early recognition of the key sources and drivers of technology trends, organizational leaders can be equipped to wisely deploy their resources. *Techonomics* examines four foundations of healthy organizations: energy, communication, computation, and community. A method of tracking market progress based on measuring both technology performance and economic cost provides a tool to consistently monitor advancement of any endeavor. Three contemporary trends based on electronic advancement (Moore's Law), network expansion (Metcalf's Law), and increasing productivity (Coase-Downes-Mui Law) are forcefully driving organizations in the twenty-first century. Examples of successful companies (Dell, Wal-Mart, etc.) utilizing emerging operational business models are revealed and explained.

Adam Smith's laws of supply and demand are being challenged by a world productive capacity that can overproduce manufactured goods and create infinite supplies of information. *Techonomics* points to the dawning of the Virtual Age, where continually increasing productivity creates more output with less labor and resources in the most effective organizations.

Author

H. Lee Martin, Ph.D., is an engineer, entrepreneur, teacher, and innovator. He holds 19 U.S. and international patents, has published dozens of technical journal articles, cofounded the now public company iPIX, and was selected as the National Young Engineer of the Year in 1988. He has earned two *Research and Development Magazine* R&D 100 Awards for innovative products in robotics and remote imaging and the NASA Small Business Innovative Research Technology of the Year Award. After earning degrees from the University of Tennessee and Purdue University, Dr. Martin developed remote robotic systems for the Oak Ridge National Laboratory. He founded TeleRobotics International, Inc., in 1986, which went public as iPIX in 1999. Throughout his professional journey, the combined impact of technology and economics shaped his understanding of trends in business and society, resulting in the development of the concepts described in *Techonomics*.

Acknowledgments

The key concepts for *Techonomics* grew out of a Fall 2004 article written for the *Bent* of Tau Beta Pi, the publication of the world's largest engineering honor society. I am most grateful to Jim Froula, the executive director of Tau Beta Pi, who provided this opportunity to introduce the concepts of Techonomics in print. Appreciation is also extended to David Keim, business editor, and Larissa Brass, business writer for the *Knoxville News Sentinel* for providing the opportunity to publically explore Techonomics by observing developments in local companies. These two writing endeavors provided a platform from which to create this book. Special thanks to Dr. Adedeji B. Badiru, the Head of the University of Tennessee Industrial and Information Engineering Department, for encouraging the CRC Press to consider this book and guiding me through the publishing process. It is one thing to see a company from conception to the public markets in the midst of the Internet boom; it is quite another to have the opportunity to describe the process with feedback from area readers to hone the lessons learned. For this opportunity, I sincerely thank these people.

I have been blessed with a highly skilled and superlatively cooperative team in the preparation of *Techonomics*. Dr. Russel Hirst provided extensive technical editing and contributed many ideas to the text. Joseph Nother added creativity and the professional touch to all the graphics herein. Colleen Steiner patiently and thoroughly typed, edited, referenced, acquired rights, and formatted the book through many revisions and was consistently thorough in her efforts. I owe the quality of the final product to their untiring efforts and the efforts of the team at CRC Press under the direction of Ms. Cindy Carelli.

Many eyes from different walks of life read and commented on the drafts. These folks contributed unique insights from their life experiences and the end product is better from their input. Technologists Dr. Adedeji B. Badiru, Daniel P. Kuban, Larry Perry, Martha Polston, and Vig Sherrill; businessmen Clint Butler, Dan Casey, David Coffey, Dr. Cary Collins, Burt Rosen, Jeff Kuban, and Scott Hutchinson; theologians Dr. Doug Banister, Brad Brinson, and Dr. Chris Stephens; and family Carla and Nina Martin contributed their vantage points on the concepts, approach, and opportunities for improvement. I am thankful for their sharing of time, wisdom, and friendship in this effort. The encouragement and mentoring of Larry Perry has been and always will be foundational to my efforts. Thank you all.

Contributor

Dr. Russel Hirst, associate professor of English and director of the Program in Technical Communication at the University of Tennessee, enriched *Techonomics* by providing a thorough edit of the entire text, including substantive contribution throughout.

Dr. Russel Hirst
Associate Professor of English
Director, Program in Technical Communication
Department of English
The University of Tennessee
401-A McClung Tower
Knoxville, TN 37996-0430

Table of Contents

SECTION 1 A Techonomic Primer

Chapter 1 Introduction to Techonomics 3
Introduction 3
Goal of this Book 4
From Biology to Business 5
Techonomics: The Definition 10
Fundamental Assumptions 12
A Leading Indicator: The Military 13
Summary 15
      Purpose of Techonomics 15
      Key Analogy 16
      Key Techonomic Assumptions 16
      Key Terms 16
Questions 16
References 17

Chapter 2 Seeing the World through Transactions 19
Story of Ronald Coase 19
Transaction Cost Analysis: The Make-or-Buy Decision 20
Hidden Transaction Costs 21
The Importance of "Perfect Information" 22
Defining Techonomic Metrics 24
Techonomic Metric Process 26
Summary 26
      Transaction Cost Analysis 26
      Contributors to Transaction Costs 26
      The Techonomic Metric 27
      Key Terms 27
Questions 27
References 28

SECTION 2   A Techonomic Perspective of History

Chapter 3   Organizational Evolution Resulting from Technological
Advancement: A Timeline 31

Introduction..31
A Timeline of Technology..32
The Four-Square Principle: Organisms/Individuals...................................36
The Four-Square Principle: Organizations/Society....................................36
    Physical — Energy 37
    Mental — Computation 39
    Social — Communication...39
    Spiritual — Community..40
Summary...41
    Technology Timeline...41
    The Four-Square Principle for Individuals.......................................41
    The Four-Square Principle for Organizations 42
    Key Terms 42
Questions..42
References..43

Chapter 4   Creating Techonomic Metrics..45

Introduction..45
The Techonomic Sweetspot...46
An Example: Digital Photography Techonomic Metric 48
    Digital Camera Techonomic Metric 48
Military: Technology Advance without Economic Constraint 51
    Military TM: Historic 51
Energy: Side 1 of the Organizational Square 54
    Energy TM: Historic 55
    Energy TM: Comparative 56
Computation: Side 2 of the Organizational Square..................................57
    Computation TM: Historic..58
    Computation TM: Comparative 60
Communications: Side 3 of the Organizational Square 64
    Communication TM: Historic...65
    Communication TM: Comparative 67
Community: Side 4 of the Organizational Square....................................70
    Techonomics of Sustainability..71
    Agriculture 73
        Agricultural TM — Historic..73
        Agricultural TM — Comparative 74
    Health and Environment 74
        Community Health TM — Historic 75

    Community Health TM — Comparative 76
  Media/Entertainment 77
    Community Mass Media Influence TM — Historic 77
    Community Mass Media Influence TM — Comparative 78
  Religion or "Spiritual" Condition 79
Reflections on Interdependence 79
Summary 81
  Techonomic Metric 81
  Leading Indicator 81
  Energy 81
  Computation 81
  Communication 81
  Community 81
  Key Terms 82
Questions 82
References 83

SECTION 3 *Techonomics at the Turn of the Twenty-First Century*

Chapter 5 The First Three Laws of Twenty-First-Century Techonomics 87

Introduction 87
Moore's Law: Ubiquitous Computing 88
  First Law of Techonomics Adapted from Moore's Law
    (Law of Ubiquitous Computing) 88
  First Law Implications 90
Metcalfe's Law: Ubiquitous Global Network 91
  Second Law of Techonomics Adapted from Metcalfe's Law
    (Law of the Ubiquitous Global Network) 91
  Second Law Implications 92
Coase-Downes-Mui Law: Diminishing Organization Size 93
  Third Law of Techonomics: Coase-Downes-Mui Law
    (Law of Diminishing Organization Size) 93
  Third Law Implications 96
The Franchise Effect: Growth through Replication 96
Summary 98
  Three Fundamental Laws of Techonomics 98
  First Law — Modified Moore's Law
    (Law of Computational Ubiquity) 98
  Second Law — Modified Metcalf's Law
    (Law of the Expanding Global Network) 99
  Third Law — Coase-Downes-Mui Law
    (Law of Diminishing Organizational Size) 99

The Franchise Effect 99
    Key Terms 99
Questions.. 99
References... 101

Chapter 6   Emerging Twenty-First-Century Techonomic
                Business Models 103

Introduction... 103
Positive Cash-Flow Manufacturing: Dell 104
    Endeavor Cash Float TM 108
Positive Cash-Flow Retail Distribution: Wal-Mart 109
    Competitive Margin TM 113
Debtless Facility Expansion: Walgreens 114
    Customer Value TM 115
Predictable Antiquation: Intel.. 116
    Predictable Obsolescence TM 118
Business at the Speed of Light: Microsoft 119
Virtual Retail: Amazon... 120
    Free Cash Flow TM 120
Virtual Reselling: eBay 123
    Transaction Efficiency TM 124
Virtual Media: Apple.. 125
Emerging Techonomic Conclusions 127
Summary 128
    Emerging Organizational Processes 128
    Positive Cash-Flow Manufacturing and Retailing....................... 128
    Predictable Antiquation 129
    Virtual Operations 129
    Selling Hardware as Portal to Media Sales................................. 129
    Financial Management Includes the Time Management of
        the Transaction 129
    Key Terms 129
Questions.. 129
References.. 131

Chapter 7   Emerging Techonomic Trends 133

Introduction... 133
Energy: Journey to Renewable Energy Resources 133
    Biological Sources 134
    Cyclical Sources... 134
        Wind Sources 135
        Wave Sources 135
        Solar Voltaic Sources 135

  Chemical Sources 136
  Nuclear Sources 137
  Contemporary Energy Technologies 138
Computation: All Things Digital 140
Communications: Expanding Control and Influence 142
Community: Increasing Efficiency from Specialization Yields
  Increasing Interdependence 147
  Protectionism 148
  Entitlement 150
Summary 151
  Twenty-First-Century Techonomics 151
  From Innovation to Trend 151
  Energy: Journey to Renewable Energy Resources 151
  Computation: All Things Digital 152
  Communications: Expanding Control and Influence 152
  Community: Increasing Efficiency from Specialization Yields
    Increasing Interdependence 152
  Key Terms 152
Questions 152
References 154

SECTION 4 Postindustrial Challenges and Techonomic Answers

Chapter 8 Techonomic Market Crises and Recommendations 157
Techonomics Natural Selection Mechanism: Competition 157
Energy: Economic Reason or Ruin 160
Healthcare: Inverted Techonomics and Its Implications 163
Education: Techonomics of Monopoly 167
  K to 12 Education, Public and Private 168
  Higher Education 171
Government: Techonomic Effect in Macroeconomics 174
Summary 179
  Competition Is the Techonomic Equivalent of Natural Selection 179
  Organizational Evolution Fails when Competition Is Removed 179
  Three Domestic Markets Defy Techonomic Trends 179
  Energy Production Economically Inhibited by Domestic Regulations 179
  Healthcare System Lacks Personal Responsibility 179
  Education System Perpetuates a Failing Monopoly 180
  Government Accounting Adjustments Mask Performance 180
  Key Terms 180
Questions 180
References 181

Chapter 9 The Techonomic Future 183
Expanding the Boundaries 183
    The Expanding Third Dimension 183
        Computational Cube or New Platform? 184
        Vertical Growing and Living 184
        Three-Dimensional Entertainment 185
        Three-Dimensional Printing 187
    Getting Really Small 187
        Space: Nanotechnology 188
        Time: Actions in Femtoseconds 189
    Extremes of Temperature: Energy Hot and Cold 189
    Biological Processes 191
Vanishing into the Virtual 192
From Adam Smith to Techonomics 196
The Techonomic Worldview 197
    Capitalism vs. Socialism 198
    Free Market vs. Protectionism 199
    Globalism vs. Nationalism 199
    Personal Responsibility vs. Entitlement 200
    Merit vs. Diversity 200
    The Invisible Hand Leading Organizational Evolution 201
Summary 202
    Techonomics' Invisible Hand 202
    Techonomic Evolution 202
    The Virtual Age 202
    Key Terms 203
Questions 203
References 204

Afterword 205

Appendix 1   Terminology Related to Techonomics 209

Appendix 2   Example of Process for Developing a Techonomic Metric 217

Index 219

Dedication

*To my children –
Nina, Ashleigh & Daniel
I dare you
to make a worthy contribution to the future*

Section 1

A Techonomic Primer

1 Introduction to Techonomics

tech*o*nom*ics (těk' ě -nŏm'ĭks) n.- The study of trends in technology and their resulting economic effects on organizations.

INTRODUCTION

It only takes one thing to succeed —

Good judgment.

It only takes one thing to get good judgment —

Experience.

It only takes one thing to get experience —

Bad judgment!

The adage above is comical because it rings true. You can do everything yourself, learn all your own lessons, make all your own mistakes, or you can learn from the experience of others and benefit from their mistakes. Besides learning from other's experiences, one of the best ways to avoid mistakes is to understand the impact of trends. *Trends are to a business as trade winds to a ship: it makes a world of difference whether they are with you or against you.* The experienced seek to understand and harness them. It matters not whether your "business" is investing, guiding a family, advising a student, leading a nonprofit, or deploying limited resources for a company: understanding the trends shaping the present and future of the world around you will benefit your decision making. Techonomics, the study of trends in technology and their resulting economic effects on organizations, is offered as a tool for understanding the rapidly changing world and providing advantageous insight for leaders. It will give you a thought process, a way to help you spot and analyze technology trends to aid decisions that are with, not against, the winds of change.

Much of what we regard as "intelligence" can accurately be described as pattern recognition: the shape of a diseased cell, the color of the sky indicating approaching weather, the cycles of the business market. Specific, measurable advances in technology are patterns for economic and organizational change. By following major changes in technology as a leading indicator, you can anticipate many economic and cultural developments. The theory of techonomics is a worldview perspective. Techonomics provides methods to consistently track technology developments and advantageously anticipate resulting organizational and economic changes.

Techonomics: Theory of Industrial Evolution provides a simple framework to observe, describe, analyze, and *predict* organizational changes by methodically tracking technological advancement. This cause-and-effect relationship between technological advance and economic progress allows us to anticipate societal trends. *Techonomics transforms the old adage "Follow the money!" into "Follow the technology!"* Understanding the economic implications of the effective use of technology improves discernment for the wise deployment of resources. Techonomic analysis provides organizations a far greater chance of adapting, surviving, and thriving in today's globally competitive environment.

GOAL OF THIS BOOK

Perspective! The goal of this book is to give you a techonomic perspective for observing and understanding our rapidly changing world. ***A fundamental awareness and working knowledge of techonomics will enable you to anticipate market trends that will affect your future at home, in your community, and in your profession.*** The ability to apply techonomics to your endeavors will develop as you consider the examples presented here and repeatedly consider techonomic trends as you deploy the resources of your organization.

Figure 1.1 shows the cycle of learning. It is a broadly applicable concept illustrating the stages of acquiring any skill. Think about the cycle in terms of a skill most of us have acquired: riding a bicycle. At two, you do not know what a bicycle is, and you do not care that you do not know. The two-year-old future bicycle rider is in Quadrant I of the chart. At about four, you realize bicycles exist, but you are a bit scared of getting onto one without guidance. You do not have one, anyway! The four-year-old is in Quadrant II, and the learning process is set to begin, because

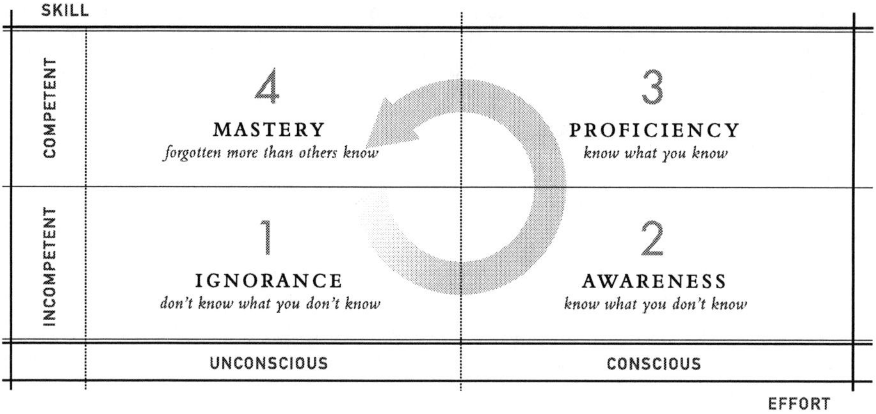

SKILL		
COMPETENT	4 MASTERY *forgotten more than others know*	3 PROFICIENCY *know what you know*
INCOMPETENT	1 IGNORANCE *don't know what you don't know*	2 AWARENESS *know what you don't know*
	UNCONSCIOUS	CONSCIOUS

EFFORT

FIGURE 1.1 The cycle of learning.

the *desire* for knowledge has entered the learner's mind. He knows what he does not know — how to ride — and he does not have access to a bicycle. But he desires the knowledge because of the freedom or enjoyment riding a bicycle appears to offer.

The six-year-old gets his or her first bicycle, complete with training wheels, and with sufficient desire, promptly moves into Quadrant III. The process of training the muscles, mind, nerves, and equilibrium begin in earnest. The length of time required to master the skill in Quadrant III depends on many things. In our bicycle example, these inputs to mastery might include the amount of practice, the basic motor skills of the child, the terrain, the ability to overcome fear of falling off, etc. For most, Quadrant IV is reached in a season or two, and from that point forward, one forever knows how to ride a bicycle, even if the intricacy of how they learned is long forgotten. The skill is mastered.

Do you desire to comprehend the trends of technology and their economic impacts to better aide the organizations you serve? Since you have purchased this book — or are at least investing time in it, however you obtained it — you evidently possess this fundamental motivation. Before today, you probably had no idea what techonomics was. The word and its definition were created to give form and substance to the ideas that have now been delineated in this book. You started in Quadrant I. While reading this introduction you are crossing into Quadrant II: awareness of techonomics. You are deciding whether or not the investment of time in learning more about this subject will be rewarded by the value this knowledge will provide. The remainder of this book reveals the cause-and-effect relationship between technology and economics on the evolution of successful organizations. ***You will become aware of continuing patterns, instigated by innovation and disseminated by commerce, that forcefully shape all vibrant organizations.***

Your techonomic advantage is the ability to discern which technology trends in your chosen field of endeavor will result in economic value to your organization, and methods for determining the proper timing for deployment of new technologies. This techonomic insight moves you from awareness (Quadrant II) to application (Quadrant III). As you apply techonomic principles to wisely deploy valuable resources for the benefit of your organization, you will reach Quadrant IV: mastery. It is a journey worth taking to discover an invaluable worldview for the mind: techonomics, the theory of organizational evolution.

FROM BIOLOGY TO BUSINESS

My grandfather grew up across the classroom from the folks he would compete with for life.

My father grew up across the country from the folks he would compete with for life.

I grew up across the world from the people I would compete with for life — only they knew it and I did not!

In 1859, Charles Darwin published his book *On the Origin of Species by Means of Natural Selection*, or the *Preservation of Favoured Races in the Struggle for Life*, popularly known as *The Origin of Species*.[1] This book was the culmination of a lifetime of observations, most of them taken during a voyage around the world from 1831 to 1836. The voyage included a pristine venue, the Galapagos Islands.

His framework for organic evolution was based on what he observed about mutation, variation, and adaptation to environmental conditions. He developed the concept of natural selection, commonly known (after Herbert Spencer coined the phrase) as "survival of the fittest." The idea was, organisms that developed various features differing from the rest of the population tended to prosper and survive and reproduce if those different features — even slightly different — gave them some sort of advantage in the environment. For example, a coloration that blended in a bit better with the surroundings could better disguise an organism from predators, giving the organism a better chance of survival. These traits would then be passed down to offspring, and in some of those offspring, advantageous traits might become even more pronounced.

This theory has been widely taught as a structural explanation for the diversity of living organisms. It has also touched off countless ideas and debates, of all kinds. In the modern age, scientists and theologians attack, champion, or modify aspects of Darwin's theory to explain the origin and development of life, but the value of an organized theoretical framework remains. Without passing judgment on its factual validity, **we can use Darwin's Theory of Organic Evolution as a way to view the interplay between key components in a complex, changing system and, by analogy, understand organizational evolution through techonomic progress.** This parallel extends the organic to the organizational, the biological to the industrial, Darwinism to techonomics.

Organic evolution, as described by Darwin, is a process by which organisms change by mutation during procreation and live or die by natural selection, creating a stronger successive generation. This theory provides a framework in which to study, classify, and analyze the diversity of biological life. The techonomic theory of organizational evolution describes the technological mutation, economic adaptation, and growth of organizations. In many ways, the success or failure of human organizations parallel the success or failure of individual living organisms. The similarities between the processes that drive these two theories will be frequently used as a framework for understanding the new relative to the familiar. Both approaches are theories, but they provide a framework for observing patterns in the changing worlds of life and organizations.

The theory of techonomics is based on personal observations spanning two decades of business development in a unique period: the dawning of the Internet. Techonomics views the organization as the evolving "organism." Unceasingly, throughout history, organizations are born, grow, mature, and ultimately die. Technology is the driving force that causes change, or "mutations," in how organizations function. The competitive economy is the "environment" that imposes "natural selection" on organizations based in no small part on their effective adoption of technology. Successful organizations grow to "procreate" again. Unsuccessful organizations ultimately become economically bankrupt and "die." **During the emergent**

Introduction to Techonomics

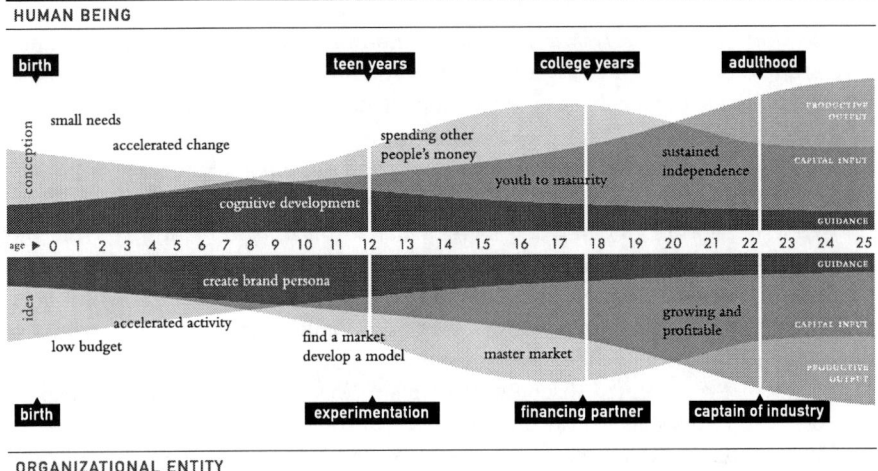

FIGURE 1.2 Parallel between the life of an organization and the life cycle of a human being.

period of the Internet, the entire business life cycle was rapidly accelerated, providing a unique, historical opportunity to observe the forces that shape organizations. Once viewing organizations as living organisms that must adapt to changing conditions, the forces that cause mutation or change can be monitored and charted.

In the U.S. in the year 2000, the number of businesses increased by over 500,000.[2] While some initiated new operations, others declared bankruptcy or ceased operations. Figure 1.2 shows the parallel between the life cycle of an organization and the life cycle of an individual person. Both entities, organization and individual, begin very small at a definable instant. The organization starts as an individual's conceptual idea, the person starts with physical conception. At this stage, both are fragile, not well defined, and must be nurtured to grow. If gestation continues, birth follows. Now the entity becomes recognized by the world: the organization files a legal charter and is named, the child is born and given a name and birth record.

In the early years of life, for both the organization and the person, a tremendous learning curve and physical development must take place. Size and financial requirements are small, but the need for guidance, protection, and training are large. The experimentation and exploration continue for a season of life until both the organization and the person find direction and focus. At this time, financial requirements may increase, as resources are needed to pursue opportunities. Business organizations need growth capital for manufacturing, inventory, sales force, etc.; individuals need support for things like education, shelter, and transportation as they try to establish their independence. As maturity comes, both organizations and individuals reach a plateau of sustenance, knowing their roles and performing those roles in exchange for the resources needed to thrive.

In the course of time, both organizations and individuals pass away, giving rise to the next generation. This next generation may look nothing like the previous one in composition and purpose. *The life-cycle time frame for any entity, organization, or individual varies greatly, but the idea of generational cycles for a typical lifespan is relevant to the concept of organizational evolution.* Certainly, throughout history organizations have been born, grown, flourished, and died giving way to a next generation more prepared for the demands of the marketplace. Understanding and describing the factors contributing to the evolution of human organizations through the ages provides the motivation for the "Theory of Organizational Evolution."

While the norm is a continual, incremental advance of technology — and this parallels Darwin's notions about "gradualism" in evolution — there sometimes occurs a discontinuous leap of innovation that powerfully impacts the world in an irreversible way. For example, the discovery of atomic energy, the introduction of mass production methods, the development of the microprocessor, and the creation of the Internet represented discontinuous innovations that had significant societal impacts beyond the realm of technology.

Big, seemingly discontinuous advances like these parallel something evolutionary theorists call "punctuated equilibrium," a variation/challenge to Darwin's theory that tries to account for seemingly big leaps of development in the fossil record. It is a concept used to describe changes for which scientists cannot find "gradual" explanations. But techonomics is an observable theory. Even big, discontinuous technological innovations do not happen in isolation from the economic pressures all organizations must endure. They occur within the bounds of science, technology, economics, and society. The competitive environment of the marketplace ultimately determines successful innovations that live to see another day.

Survival of the fittest in the domain of the organization is determined by the economic survival of the organization. Organizations that embrace technology as a means to continuously improve efficiency maintain and extend their competitive position. An organization that becomes insular to technological advancement eventually ceases to be competitive and ultimately fails economically. Therein lies the generational cycle of organizational birth, growth, and death that mirrors the same pattern in the living organism.

In Darwin's theory of organic evolution, generation upon generation of organisms mutate, adapt, and procreate, leading to diversity and advancement in the living species. The techonomic theory of organizational evolution observes that the technologies successful in the marketplace are rapidly spread to other organizations. Adoption of best practices is one indicator of a healthy organization. In both theories, the sequence of events occurs over time: first the change caused by mutation (technology advancement) and then the determination of success or failure as a result of environmental (economic) natural selection. Table 1.1 shows the parallels between evolutionary frameworks. This table lists the elements of the analogy between Darwin's Theory of Organic Evolution and the Techonomic Theory of Organizational Evolution.

In the theories, both evolutionary processes continue through the ages. The cycles (birth, growth, and demise) of living organisms and thriving organizations continue relentlessly in our world. The next two figures provide a parallel visual representation of Darwin's theory for organisms and Techonomic theory for organizations. Figure 1.3 shows the key process relationships in Darwin's Theory of Organic Evolution.

Introduction to Techonomics

TABLE 1.1
Parallels between Evolutionary Frameworks

	Darwin's Organic Evolution	Techonomics Organizational Evolution
Field	biological organism	human organizations
Change agent	mutation	technological advance
Playing field	environment	economy
Success filter	natural selection	competition
Survivor	survival of the fittest	survival of the efficient/effective
Options	reproduction or death	perpetuation or bankruptcy
Motivation	competition for survival	free market economic competition
Survival keys	food, shelter, protection	capital, quality, efficiency

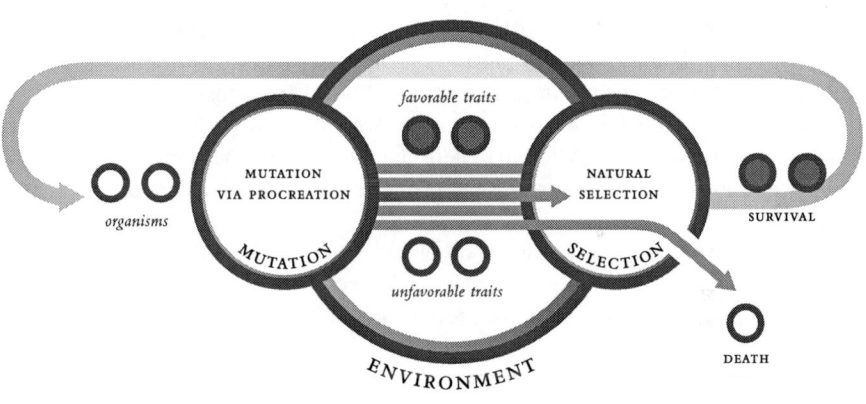

FIGURE 1.3 Key process relationships in Darwin's Theory of Organic Evolution.

Figure 1.4 shows the key process relationships in the Techonomic Theory of Organizational Evolution. The analogy between the two theories and their key elements is evident. Deeper analysis will reveal that analogous forces act to advance both systems.

Consider the fundamental importance of competition to both organisms and organizations. When the organism is removed from its natural habitat and placed within the protective environment of a zoo, its developmental patterns change. No longer is it required to "fend for a living," as it is protected from its natural enemies and its food and shelter are provided. Most zoologists will concur a carnivorous animal raised within a protected environment will quickly perish if turned back to the wild without extensive retraining. The organization is no different. ***The competitive pressures of the free market ultimately force businesses to adopt best practices***

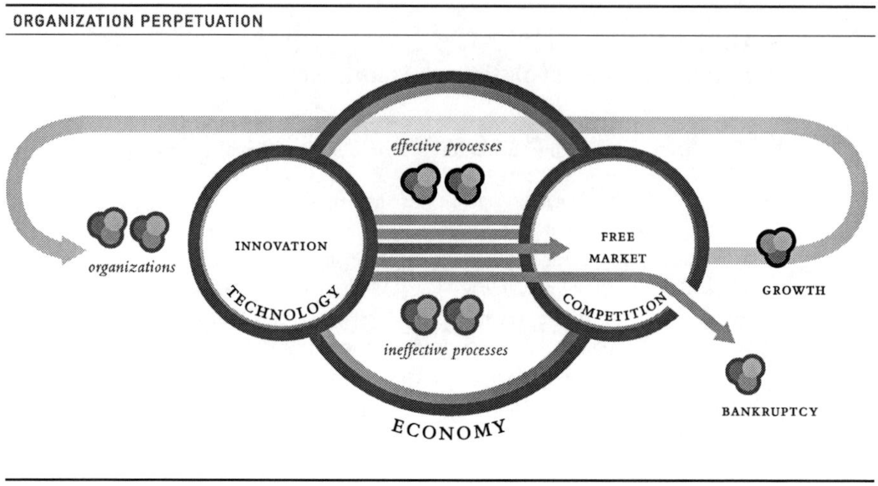

ORGANIZATION PERPETUATION

FIGURE 1.4 Key process relationship in Techonomic Theory of Organizational Evolution.

or face elimination by others. In a protected market or a government-constrained economy (much like the zoo), organizations are not forced to adopt best practices. Over time, these organizations steadily fall behind their competitive counterparts. Ultimately, they forget how to compete and continue to exist only as long as protections are in place. After a generation or two, the removal of protective barriers to the competitive marketplace results in massive organizational upheaval or cessation.

The most graphic illustration of this premise occurred with the falling of the Berlin Wall and the stark economic contrast observed between the former socialistic economies and other free market economies of the industrialized world. The economic contrast in East and West Berlin, in North and South Korea, or in the struggle for the former states of the Soviet Union to compete in the global economy gives tangible evidence of the "natural selection" process resulting from different economic structures. Overnight, organizations that had been protected by the umbrella of monopoly or centralized planning found themselves forced to compete for survival in a demanding economic environment. Many quickly perished. These nations and their productive outputs during this era provide clear evidence of the disparate results from various economic systems that either promote organizational evolution via techonomics or constrain it.

TECHONOMICS: THE DEFINITION

tech*o*nom*ics (tĕk′ ĕ -nŏm′ĭks) *n.*

1. The study of trends in technology and their resulting economic effects on organizations.
2. A theory of organizational evolution that results from technology advance selected by economic success.

Introduction to Techonomics

Combination of technology + economics:

- **technology:** Greek *tekhnologiā*, systematic treatment of an art or craft: *tekhnē*, skill or craft + *-logos*, knowledge or reason.[3]
- **economics**: Middle English *yconomye*, management of a household, from Latin *oeconomia*, from Greek *oikonomiā*, from *oikonomos*, manager of a household: *oikos*, house + *nemein*, to allot, manage.[4]

The two words, *tech*nology and ec*onomics,* have been fused to derive *techonomics:* the study of how technology shapes organizations through the economy. For clarity in the engineering professional, techonomics should not be confused with *Technometrics,* which is the name of an industrial engineering technical publication focusing on measurement of technology. ***Techonomics combines technology and economic measurements to objectively observe trends related to organizational success.***

Among Sir Isaac Newton's famous statements is this one: *For every action there is an equal and opposite reaction.* Techonomics views technology as the driving *action,* the instigator, of new opportunities, the force behind new organizing principles. The economic *reaction* of organizations is the observable result. The reactions in organizations are not precisely "opposite," of course, but measurable economic costs for technology implementation are fundamental to predicting the likelihood of successful deployment. ***Technological advance creates opportunities for new organizational structures, while economic success determines long-term viability of organizations.***

Traditionally, economics has been guided by the study of fundamentals of supply and demand in the free market to determine the pricing of goods and services.[5] Marketplace fundamentals of supply and demand are no longer the same in an era where technology makes it possible for most consumer goods to be overproduced. Even more telling, constraints on "supply" are nonexistent in the world of a "virtual" information-based product. As Nicholas Negroponte describes in *Being Digital,* we are well into the economic journey from atoms (things) to bits (information).[6] The fittest organizations in today's business environment survive and prosper because they focus on deploying *all* resources wisely. Successful organizations capitalize on the understanding that technological advancement precedes economic opportunity.

The lexicon used to describe commerce is filled with an ever-expanding list of new terms: mass customization, globalization, networking, off-shoring, virtual companies, rationalization, micro-multinational, rolling warehouse, just-in-time. It is difficult to sort all these changes out, let alone put them all back together into a worldview that makes sense. Techonomics provides an observation post — an analysis process grounded in technology — for evaluating the root causes producing these emerging terms and practices. It reveals the *why* behind the *what,* fueling major trends in fast-paced twenty-first-century organizations.

A few common terms will be used frequently in the discussion of techonomics. The specific techonomic context of these terms will aid your understanding of the subject. These terms include:

- *Organization* is broadly applied to reference any group of people who share a definable, common affiliation — business, geographic location, citizenship, religious affiliation, civic association, nonprofit business, volunteer group, etc. Organizations are as diverse as the people and the endeavors they pursue, yet all organizations share observable characteristics. Every organization has a purpose, some better than others. Every organization has a beginning and, probably, an ending someday — a life cycle. Every organization requires energy, direction, and communication and creates some form of community structure. For organizations to thrive, they must pursue sustainable economic models.
- *Endeavors* are the products and services that organizations produce, the valued output. Endeavors for your business organization might be products or services. For the nonprofit organization, the endeavors might be education, a catalyst for public action, or spiritual conversion. Business endeavors range from tangible products to intangible information with a myriad of possibilities. The local community might broadly classify the quality of life as its main endeavor. The societal value of endeavors relative to their cost to produce is a key determinant of the prosperity of the organization. When multiple organizations perform the same endeavor, the most efficient/effective organization typically thrives, while others diminish. In a free-market economy "natural selection" causes vibrant organizations to adopt best practices for the performance of their endeavors.
- *Transactions* are the interchange between organizations or individuals in the marketplace. Participants in transactions are either the buyer (customer) or seller (producer). Transactions are the fundamental building blocks of commerce — no transactions, no commerce. The observation of shifting organizational transaction patterns gives direct insight into the impact of technology on the economy.
- *Metrics* are the combination of measurements into a value that can be useful as an unbiased, numerical indicator of a trend when computed over spans of time. Techonomic metrics (techonometrics) combine technology measurements with economic measurements to track the market potential of innovation. A complete discussion on the formulation of techonomic metrics is found in the next chapter.

FUNDAMENTAL ASSUMPTIONS

A handful of fundamental assumptions support the applicability of techonomics to anticipating organizational trends. These assumptions, simple as they appear, are fundamental to the "natural selection" analogy that supports the relationship between technological development and economic impact. Remove any one of these four fundamentals from the marketplace and economic pressure for organizational evolution is disrupted.

- **Relatively free market** — A free market is required for the economic advantages of technological advance to be embraced. In a contrived/controlled marketplace, meritorious practices give way to political persuasion, regulation or subsidy.
- **Competition** — Competition in the marketplace is the force that drives organizations to embrace best practices. Best practices fuel the rapid pace of technology adoption, thereby advancing the economics of the enterprise.
- **Expanding knowledge** — For organizations to progress, the universe of knowledge must be expanding. The rate at which this expansion occurs in any period of time determines technology's degree of influence on organizations within the framework of the economy. Hence, in times of rapid knowledge expansion, techonomic trends become more evident and more important.
- **Self-interest** — Individuals and organizations make decisions based on their desire to survive and thrive. Without these desires, organisms/organizations — be they individual, corporate or national — cease to improve, leading to demise. While some decisions and actions are conceived out of good will, economic viability is at the forefront of entities that stand the test of time. Whether producing for the market or buying from the market, organizations expend resources on the best perceived products and services. Once self-interest for improvement is removed, by entitlement or regulation, the fundamental motivation for progress is destroyed — along with the vitality it brings.

A LEADING INDICATOR: THE MILITARY

If technology is the instigator of organizational change due to economic selection, is there an instigator of technological change? Yes. Military demands have been the driving force in the development of most technologies throughout the ages. From gunpowder to atomic energy, early digital computers to microprocessors, telegraph to Internet, semaphore to satellite, DaVinci machines to space shuttles, military needs have pushed the boundaries of science to implement practical solutions. One key reason for the military's role as the leading source of technology innovation is that military survival supersedes economic competition. Typically, such demands are accompanied by a national attitude of survival at any cost.

Technology leaps are often generated in an environment where numerous possibilities are pursued, and the end work product does not have to stand the economic pressures of the consumer marketplace. *Generous military expenditures for nascent technologies become the incubator in which ideas and possibilities become field-tested realities.* The military potential, fueled by fear of the unknown enemy rather than the economic return, becomes the motivational seed for cultivating a host of futuristic technologies. The journey from concept to proposal to prototype to early adoption traverses the most treacherous territory for any new technology, whether or not it is related to military endeavors. Many innovations do not cross this chasm for lack of financial resources, but far fewer would ever begin the journey without

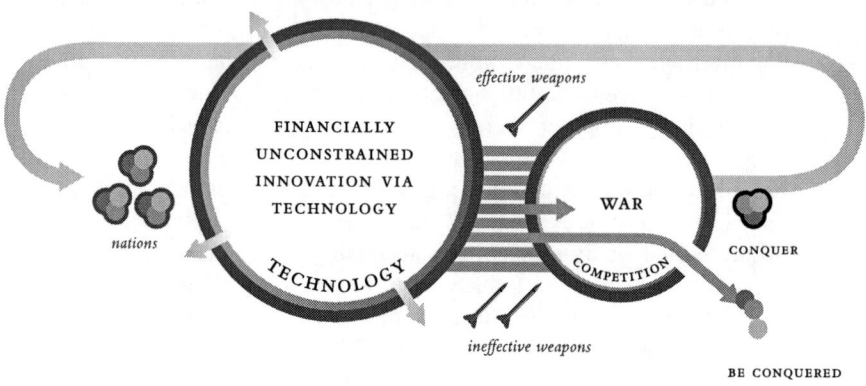

FIGURE 1.5 Elimination of economic constraints when military requirements promote technology advancement due to national survival at any cost in time of crisis.

the lifting of the economic constraints spurred by military considerations. Figure 1.5 shows the elimination of economic constraints when military needs promote technology advancement due to national survival at any cost, particularly evident in time of crisis.

This is a techonomic observation, not a judgment on the morality of this arrangement. Obviously, it would be preferable for the world to become so safe and peaceful, universally inhabited by such civilized and enlightened people, that the need for military protection could fade away. The reality is that we do have militaries and that, when they are large and well funded, the scientists and technologists working for them escape economic pressures, vastly increasing their ability to extend the boundaries of science.

Innovative research activities are not, of course, confined to the military. For example, "think-tank" environments have existed in the commercial sector, such as at Bell Labs, RCA-Sarnoff Labs, SRI, and Xerox Park. To a lesser extent, these environments exist in universities and federal laboratories. Three key ingredients supporting military technological innovation are not present in the same degree in these other organizations:

1. Research funding disconnected from financial returns
2. Organizational urgency related to national sovereignty
3. Large-scale, controlled user field deployment of promising results without the constraint of market economies driven by consumer desires and budgets

A military research effort always has an early adopter standing by; not so with private research. Bell Laboratories turned out advance after advance as long as there was a captive monopoly supporting their efforts (the Bell System). But soon after communications deregulation, the Bell Laboratories fell to the cost cutter's axe. It was sold off in pieces to other companies. Xerox Park created the foundations of the modern personal computer interface only to have its own corporation fail to grasp the value and potential of the visual operating system. Technological descendants of the Xerox Park operating system were popularized by Apple and made massively profitable by Microsoft. Even when a research laboratory supports a corporation, the risk of marketplace deployment may cause many valuable innovations to remain in the lab due to the lack of a controlled, early-adopter market. These are stages in the life of a new technology for which military applications provide the gestation: need identification, conception, theoretical development, experiment, prototype, refinement, and field trials.

Military application of technology, as a general rule, provides the bellwether of future mass applications of commercial technology. To anticipate the technologies that will emerge in the consumer market tomorrow, one need only study today's military technologies with an eye toward potential civilian applications understanding the possibilities of mass-application cost reductions. The Web, cell phones, global positioning systems, bar codes, lasers, biometrics, radio frequency identification tags — the list of currently expanding technologies derived from military beginnings is almost endless. *Although the military is not the only site of technological development, it is the premiere entity with resources, focus, and urgency to develop and field-test technologies on the cutting edge of science.* What remains for the consumer market is the shifting of technology application to civilian needs, cost reduction to meet market expectations, and improvements in operating reliability to satisfy the demands of mass applications. These things take time, even in our fast-paced Internet world. Therefore, if you watch military developments, you may just be struck by an insight about what is coming to your market in the next few years.

SUMMARY

tech*o*nom*ics (tĕk′ ĕ -nŏm′iks) *n.*

1. The study of how technology affects the economy.
2. A theory of organizational evolution that results from technology advance fueled and selected by economic success.

PURPOSE OF TECHONOMICS

- Creates a framework for spotting technology trends that will impact your organization.
- Creates a method for developing "metrics" that combine key technology attributes with economic measures. Understanding the timing of these trends results in timely deployment of resources.

Key Analogy

Techonomics is presented as a theory of organizational evolution with many parallels to Darwin's Theory of Organic Evolution (Table 1.1). Human organization is seen as a continuing process progressing through many generations of organizations, driven largely by increases in technological knowledge and its economical implementation.

Key Techonomic Assumptions

- Relatively free marketplace
- Competition
- Increasing knowledge
- Self-interest

These assumptions provide a foundation that encourages adoption of best practices between competitive organizations within the free marketplace, thereby increasing dissemination of knowledge in the self-interest of the organizations. Removing any of these "natural selection" factors weakens the impetus for organizational evolution.

Military technology development precedes civilian commercial deployment because of resources not constrained by economic returns and the captive field-deployment of early stage innovations.

Key Terms

Organic Evolution	Organizational Evolution
endeavor	transaction
metric	techonomics
competition	natural selection
self-interest	free market

QUESTIONS

1. Techonomics is about spotting trends. In your experience, what are the techonomic trends affecting personal interchange of the written word (letters, email, instant messaging, etc.)? Based on your observations of these trends, can you make any predictions about the nature of future endeavors for the U.S. Post Office? What about newspapers? Magazines?
2. The cycle of learning (Figure 1.1) represents the learning journey in acquisition of new skills. Think back on a general skill we all embraced — reading — and note the key activities/events in each quadrant. Do people or events come to mind that affected your progress? Now think of another important skill you have mastered and trace your development

and key events/people. With each skill acquired, there are many skills not developed due to lack of time or access to resources. This is the basis of opportunity costs and a key for determining how to deploy resources (time, money, labor). Think of a skill you did not pursue due to lack of time or resources (not just due to the lack of talent).
3. How can techonomics provide insight into your organization and its deployment of resources?
4. Techonomics is based on four assumptions (see chapter summary). Do you question the validity of any of these assumptions? Why or why not?
5. The Theory of Organic Evolution is a controversial subject with widely understood tenants. Is it necessary to embrace the totality of any theory in order to use its framework as part of explaining another theory? Why or why not?
6. Darwin asks us to consider a span of many millions of years in order to recognize the dynamics of organic evolution. Organizational evolution has been active since people started living together (a much shorter time span) and its effects are observable by people. What advantage does that give to the practical validation of techonomics as a theory of organizational evolution?

REFERENCES

1. Darwin, C., *On the Origin of Species by Means of Natural Selection, Or the Preservation of Favoured Races in the Struggle for Life,* Wordsworth Editions Limited, Hertfordshire, 1998.
2. O'Rourke, P.M., Total number of U.S. businesses, BizStats.com, 2003, http://www.bizstats.com/businesses.htm.
3. *The American Heritage Dictionary of the English Language,* 4th ed., Houghton Mifflin Company, 2004, Answers.com, 2005, http://www.answers.com/topic/technology.
4. *The American Heritage Dictionary of the English Language,* 4th ed., Houghton Mifflin Company, 2004, Answers.com, 2005, http://www.answers.com/topic/economics.
5. Smith, A., *Wealth of Nations,* Great Mind Series, Prometheus Books, New York, 1991.
6. Negroponte, N., *Being Digital,* 1st ed Edition, Knopf Publishing, New York, 1996.

2 Seeing the World through Transactions

Follow the money.

—Unknown

Follow the technology.

—*Techonomics*

STORY OF RONALD COASE

Dr. Ronald H. Coase was born in England in 1910. During the academic year of 1931 to 1932, he came to the U.S. on a traveling scholarship to study the structure of American industries. Coase was a socialist at the time and sought to understand capitalism. He visited Ford and General Motors. Coase came up with a puzzle: how could economists say that Lenin was wrong in thinking the Russian economy could be run like one big factory, when some very large firms in the U.S. seemed to be run very well?[1]

What came out of his enquiries was not a complete theory answering the questions he initially sought, but the introduction of a new concept into economic analysis: *transaction costs*. Coase used transaction costs to explain why there are firms. Coase wrote firms were smaller versions of the socialist approach of centrally planned economies, but they existed because of people's choices. Coase asked, why do people make these choices? The answer, wrote Coase, is "marketing costs." (Economists now use the term "transaction costs.") **Coase theorized that, if markets were costless to use, firms would not exist.** Instead, individuals would make personal, arm-length transactions. But because markets are costly to use, the most efficient production process often takes place in a firm. Coase's explanation of why firms exist birthed a body of literature on the transaction analysis. These ideas became the basis for his article "The Nature of the Firm," published in 1937 and cited by the Royal Swedish Academy of Sciences in awarding him the 1991 Alfred Nobel Memorial Prize in Economic Sciences.[2]

What Coase observed at Ford and General Motors were very large and vertically integrated manufacturing firms. Less than a generation before, Henry Ford had changed manufacturing forever by the invention of the assembly line, complete with repetitive tasks, interchangeable parts, interchangeable workers, and a product priced so the workers themselves could afford to buy it. Since Ford initiated this type of large manufacturing operation, there were not many supply sources his firm could rely on to produce parts in the quantity he needed. Part of the vertical integration resulted from necessity: no other source of large quantity supply.

Another factor supporting the monolithic, vertically integrated structure was the limited transportation and communication infrastructure. The concentrated growth of Detroit as an automotive industry center was in no small way attributable to the need for suppliers to be in close proximity to the end producer. Once suppliers of sufficient size and quality were located in the region, Ford and other growing competitors, like General Motors, began to have choices for their component supplies. They could either make the part themselves or buy it elsewhere. The firm began the evolution from a monolith to a network of suppliers. The primary driver was the cost of the part.

TRANSACTION COST ANALYSIS: THE MAKE-OR-BUY DECISION

The economic field of transaction cost analysis looks at make-or-buy decisions. The make-or-buy decision considers whether a transaction is more economically performed internally (make) or through external sources (buy). We all analyze transaction costs many times a day, often without even considering what we are doing. Sometimes the decision is obvious, sometimes not. Did you make your lunch today or buy it at a convenient restaurant? If you made your lunch, did you grow your own tomatoes? Bake your own bread? Make your own butter? You probably bought your lunch or bought the components of your "homemade" lunch, but it has not always been that way. Turn back the clock a few generations in the U.S., to the agrarian society. In that age, there were many more make than buy decisions. Living then was not, generally, as easy as now, but personal independence was much greater, because there was less reliance on external supplies for basic necessities. Observe today what happens in large cities if a power outage extends more than 2 days because of weather or system failure. Communication is limited or curtailed, pumps do not work (limiting water and fuel supplies), and food supplies begin to dwindle. Without electric power, many of our "buy" decisions are interrupted. The interdependence of modern civilization begins to fray at the edges.

Or travel to the third world today and see people taking care of their own fundamental needs. Rural economies in most of the world provide a subsistence-level existence; people live off the land. Their meager existence is balanced by considerable independence for the fundamental necessities of life. Third-world living conditions can consist of a few animals for milk, meat, and fabric; a plot of land for food; water from a well or stream; and a shelter constructed of the most basic materials. The lack of commerce (fewer transactions) due to minimal currency availability and inherent lack of corporate employment shifts the make-or-buy decision to the make side.

The make-or-buy decision determines our transportation (ride an animal, bike, bus, car, plane — or walk), our clothes (make material, sew, buy at store), our entertainment (sit and talk, sing, play an instrument, listen to others play, listen to radio, watch TV, browse the Internet), and our health (exercise, eat healthy food, sleep regularly, take pills, have operations). Economic standing, availability of items, availability of currency, needs vs. wants, and many more considerations play a role in our personal decisions whether to make or buy. Every organization (individual, family, firm, or community) decides to make or buy numerous times for every endeavor in which it engages. The organization's discernment of value in these choices often determines its long-term success. A thorough understanding of the mission, purpose, and values of the organization provides a guide to the myriad of make-or-buy decisions.

HIDDEN TRANSACTION COSTS

In an uncomplicated and trustworthy world, cost for any item would be expressed with one clear, comprehensive attribute — the price tag. You would simply ask yourself what is the best price for the item you require. Often, this is the first question considered, and if the cost is too high from external sources, the decision runs to making the product internally (i.e., the common decision to make rather than buy food in the third world).

But our world, especially the industrialized "first world," is anything but simple, and transactions of any kind can have a number of hidden costs, some of which can be devastating if not fully considered. ***These contributors to transaction costs are referred to as "hidden" because they are not as easily quantified as the price tag, even though they may be more fundamental to the ultimate value of the purchase.*** A sample of hidden transaction costs includes:

- *Availability:* Before one can buy a product, one must locate a satisfactory supplier and then set about the process of negotiating and contracting with that supplier. Does the product exist in the form desired? Will the provider sell to your enterprise? Each of these steps requires information, communication, and trust to determine if the transaction is worthy and the supplier reliable.
- *Quality:* Fitness for intended purpose has many facets, such as specification, inspection, terms, and rates of rejection, and the cost of customer returns both in actual terms and in lost future opportunities.
- *Transport:* It is one thing to buy a product, it is quite another to pay to have it moved. Some products are easily transported (software over communications systems), while others are bulky or perishable, requiring proximity in their production (for example crushed rock, medical isotopes, fresh produce). Goods in transport represent an inventory cost that impacts the total transaction cost.
- *Punctuality:* A product may be of perfect quality and inexpensive price, but delay in its delivery could close down production at a multimillion-dollar facility. Such costs must be considered as risk in the make-or-buy decision.
- *Inventory:* One must decide how making or buying a product will affect inventory. Inventory requirements are a part of the capital needed to produce and deliver products, as well as tax consequences.

- *Switching:* Switching costs (the costs associated with "switching" to another supplier) can mitigate or amplify the impact of other transaction costs. If switching costs are low (other quality sources, easily accessed, instantaneously available, etc.) then the consequences of other hidden transaction costs can be reduced because a secondary supplier can be readily obtained if the primary source fails to deliver. If the switching costs are high (sole source, proprietary formulation, unknown quality, distance), then the consequences of other hidden transaction costs are amplified.
- *Risk:* What are the organizational risks involved with outsourcing a particular endeavor? Risk could be financial (production outages due to supply interruption), strategic (lost trade secrets), or relational (conflict of interests with other partners). If the risk is too high, the financial gain from outsourcing might be overshadowed by the potential risks associated with a certain path.
- *Others:* A number of other factors — trade laws, regulations, tariffs, etc. — may determine complete transaction costs for any endeavor. Your own situational analysis will uncover these.

Figuring out many of the hidden transaction costs boils down to quality and availability of information and the trust you can place in the reliability of that information. Long-standing relationships are often worth a premium price due to the trust garnered through the years. As global competition has increased, competition in the marketplace has forced lower-priced alternatives to be considered. Likewise, the combination of the Internet and electronic commerce has allowed new sources for supplies to be located, considered, quoted, validated, surveyed, and used. **The falling cost of telecommunications has made it economical to connect distant manufacturing systems in order to monitor production rates, shipments, key quality measurements, and payments.** Today, it is as easy and potentially more cost effective to perform these tasks across the world as it is to do them across town. All these technological changes have reduced the transaction costs for outsourcing production. Monolithic manufacturing facilities are rapidly yielding to manufacturing networks that are centrally monitored, but geographically decentralized. The ideal of "perfect information" plays a key role in reducing hidden transaction costs.

THE IMPORTANCE OF "PERFECT INFORMATION"

Information has many characteristics that govern its value. Information is "perfect" when all these characteristics are not just optimized, but have reached theoretically ideal limits. Important information characteristics include:

- *Accuracy.* Accuracy determines the credibility of the information. Is it correct? If it is a measurement, is it from a working, calibrated instrument, and is the variation between measurements small enough to be inconsequential? "Perfect information" is characterized as trustworthy, precise, and actionable.

- *Timeliness.* Old information may not be actionable. Last year's holiday unit sales for personal computers may not have any relationship to the inventory requirements for this year's PC industry. If one producer bases production on last year's demand while another determines production from current online orders, the timeliness of information will greatly favor the second producer in determining production requirements. The world of just-in-time manufacturing runs on accurate and timely information. "Perfect information" is instantaneous and actionable.
- *Cost.* Many components determine the cost of information. The most significant information costs are:
  1. Origination/access cost: How much does the information cost to obtain and access? "Perfect information" is obtained at no cost.
  2. Transmission cost: How much does it cost to get the information from its source to its destination? "Perfect information" is transmitted at no cost.
  3. Searching cost: How much does searching for the information cost? Depending on how information is organized, searching can be expensive and time consuming. The advent of search engines has transformed the slow linear search into a rapid direct access approach. "Perfect information" is located at no cost.
- *Completeness.* Complete information provides the entire picture of the endeavor in question. Partial information, even when accurate, timely, and free, can leave gaps in understanding. Remember John Godfrey Saxe's poem about the six blind men believing they were describing the whole elephant?[3] As each blind man laid hold of a different part (one man felt the elephant's side and confidently declared the beast to resemble a wall, another grabbed the trunk and thought the elephant very like a snake, and so on), their perception of the elephant was completely different. "Perfect information," unlike the reports of the blind men, reveals the entire elephant. "Perfect information" is omniscient.

Consider what a great decision maker you would be with perfect information. You would have trustworthy information, instantly, at no cost, from all possible credible sources available. *In the history of information, the Internet provides a closer approximation to perfect information than has ever been available to humanity* — if only the Internet were completely credible and trustworthy! As things are, you still have to separate the wheat from the chaff. The glut of information can also be a problem as you seek information on simple endeavors and are swamped with thousands of possibilities, many of which are not valid or relevant. But the advent of the Internet, in combination with search engines like Google and Yahoo, has permanently changed the way we access information and make decisions. From linear to nonlinear, slow to instant, physical to virtual — the Internet has changed information access, guiding it toward the "perfect" ideal. Orders of magnitude more data are available today, in a timelier manner and at lower cost, than just 10 years ago. This sudden, ubiquitous availability of nearly "perfect information" is more important to the future of the fundamental organization of society at the dawn of the twenty-first century than the discovery and harnessing of electricity was to the twentieth century.

The techonomic ramifications of "perfect information" are now taking effect. *Techonomic metrics,* used in concert with perfect information, allow the tracking of trends in key endeavors. To observe techonomic effects in your endeavors, it is important to understand the process used to create a techonomic metric.

DEFINING TECHONOMIC METRICS

One key element of this book is to demonstrate how to develop a techonomic metric. ***A metric is a key measurement that is used to monitor the progress of a system.*** The amount of gasoline in an automobile is an example of a key metric frequently monitored by the driver to determine when refueling is necessary. As the fuel gauge moves from full to empty, there comes a point in time where the operator needs to take action to refuel. The timing of the refueling decision may have many other contributors (availability of refueling stations, capacity of the fuel tank, fuel efficiency of the vehicle, even the accuracy of the fuel gauge itself), but the simple, trustworthy fuel gauge is the primary metric used by the driver to decide to refuel.

Techonomic metrics for the marketplace are similar to the fuel gauge for the automobile, with the additional proviso that they measure a set of factors governing both technology (performance) and economy (cost). Continuing with the automotive analogy, a techonomic metric would divide the vehicle fuel efficiency (miles/gallon or km/l) by the cost of fuel ($/gallon or $/l) to create a techonomic metric of vehicular operating efficiency (miles/$ or km/$). Increase the fuel efficiency by better design (lighter weight, reduced aerodynamic drag, etc.) or decrease the fuel cost by using an alternative source, and the techonomic metric will increase. Techonomic progress for a single fuel type (i.e., gasoline) can be measured over time, or a techonomic comparison of the best practices for two different fuels can be simultaneously compared (i.e., hybrid-electric vs. gasoline). Good techonomic metrics provide a value that can be consistently determined by different observers combining technology and economic information into a single observable quantity. ***To qualify as "techonomic," a metric must include data for both performance (technology) and price (economy) into a single quantity representing key elements of the endeavor being monitored.*** A more detailed description of how to develop a techonomic metric can be found in Chapter 4 and Appendix 2.

Understanding any marketplace requires observation of the key transactions related to endeavors. ***Transactions expose the societal value of endeavors allowing their observation.*** Once the significant contributors to the performance and cost of endeavors are determined, the next step is to consider how shifts in technology affect those key contributors. Technology may affect production, distribution, source location, financial exchange, publicity, shelf life of goods, or a host of other factors that ultimately determine the quality/cost proposition provided to the end user.

In bullet form, here is the simplified thought process for developing a techonomic metric for a give marketplace endeavor:

- Find the technology and economic components required to deliver the fundamental endeavor in the market and combine them into a techonomic metric.
- Gather historic market data for the technology and economic components, paying close attention to any major shifts relative to technology introductions that have affected the techonomic metric. Where historical data in terms of dollars are not available, consider the use of a timeless economic equivalent such as labor hours, land mass, or production speed.
- Project the effect of emerging technologies or economies of scale on the future of the techonomic metric.
- Visualize key technological opportunities that could restructure the techonomic metric for a given endeavor.

This process seeks insight into the timing and magnitude of the possibilities as well as the magnitude. Once we have established the techonomic metric established for a given endeavor, the relationship between technology advance and economic results is defined. But it is also critical to understand that the time required for the technology to impact the marketplace is critical to the wise deployment of resources. How many failed entrepreneurs lament that they were 5 years ahead of their time? The desire is to be 5 minutes ahead of your time, utilizing new technologies to master a receptive marketplace. Advances in mass communications have increasingly made the marketplace adoption of technological developments more rapid. Often, the determining factor of the market readiness of a new technology is the cost for widespread deployment.

The cost acceptable to the marketplace may not be the initial cost that a supplier can offer profitably. As a product gains market share, economies of scale in production and distribution enable cost reductions in the product that were never imagined by the original developers. In *Crossing the Chasm,* Geoffrey A. Moore elegantly describes the reducing cost curve for new product introductions as they find an ever-increasing market.[4] Increasing market presence allows mass production techniques and spreading of development costs over a larger product base, thereby reducing the product price.

The reduction of production costs is most evident in the field of electronics. An embedded (self-contained) control system for the first generation of a product may be cost prohibitive but required to become the first to market. A loss-leader price may be required to capture the initial marketplace, but the combination of quantity production and progressively diminishing electronic costs may ultimately reward the successful first mover as the marketplace embraces the product.

Before finishing our discussion, let us note ***techonomic metrics can be used for comparison of different categories of information depending on the insight sought.*** Metrics may be applied to simultaneous operations at different locations or using different processes to discover and understand the best approach for an endeavor. Metrics may be applied to the same endeavor at different points in time to determine how technology changes have affected the endeavor or might affect it in the future.

In the end, the metric is only as good as its components and the data used in its calculation, but the value of including functional and financial information in an overarching measurement is the key of techonomics. This framework provides a point of observation giving perspective and insight into diverse and complex decisions.

TECHONOMIC METRIC PROCESS

- Determine major technical contributors to performing an endeavor.
- Determine a consistent cost measurement for the delivery of the endeavor.
- Combine the technology performance with the delivered endeavor cost to create a single metric (typically: unit/$).
- Track techonomic metric over time to monitor evolution of the endeavor.
- Techonomic metrics are most valuable when:
  1. They broadly address all important contributors.
  2. Subjective data are minimized.

In periods when technological change is slow, market forces of supply and demand control the economy. But in periods of rapid development, technology revolutionizes production methods, resulting in enormous economic efficiencies and organizational impact. The fundamental purpose of techonomic analysis is to be able to spot trends that lead to success in the marketplace and also use understanding of those trends to determine when to enter the marketplace. In the following chapters we will use history as a teacher of techonomic trends and develop techonomic metrics to predict those trends into the future.

SUMMARY

Transaction Cost Analysis

Dr. Ronald Coase initiated a method of economic analysis called "transaction analysis" as he tried to understand how companies decide which activities they should accomplish in house and which they should buy from external suppliers. Transactional analysis has come to be known simply as the "make-or-buy" decision: do we make whatever is needed inside the company, or do we find a supplier to buy it from?

Contributors to Transaction Costs

While the most obvious transaction cost is the monetary price for a good or service, there are several other hidden costs that may bear heavily on the make-or-buy decision. These include, but are not limited to: availability, quality, consistency, punctuality (of delivery), transport, switching cost, and risk of supply interruption. While cost may be a comparative data point between options, the hidden costs are often "make or break" in determining a possible outsourcing option. There is no substitute for relational trust in minimizing the impact of hidden costs in the transaction analysis equation, but low switching costs (easily accessed replacement sources) reduce the risk of a mistake.

THE TECHONOMIC METRIC

Combining a technology performance measurement with its underlying cost creates a techonomic metric. Techonomic metrics provide a value to track key endeavors in an industry or market by measuring the contributors and calculating the metric as time passes. Techonomic metrics provide insight into major technology developments that significantly affect performance or cost, leading to new market penetration or competitive advantages:

- Sum major contributors to an endeavor's cost.
- Observe differences in performance or costs due to technology advancement.
- Broadly address all major contributors.
- Minimize subjective data.

KEY TERMS

transaction costs	make or buy decision
vertical integration	hidden costs
switching costs	perfect information
techonomic metric	

QUESTIONS

1. Ronald Coase's transaction cost analysis approach is applicable to many of our daily endeavors on a personal or corporate basis. Consider the make-or-buy decision for something you regularly do for yourself (mow grass, wash clothing, cook breakfast, etc.) What contributor(s) would have to change in the "make-or-buy" decision for you to decide to buy that product/service rather than make it yourself?
2. Your neighborhood probably has several grocery stores. What are some of the possible differentiators that determine which store you frequent? Analyze why you shop at one more frequently than another, considering as many factors as possible in the make-or-buy analysis.
3. Restaurants are a very competitive business. Analyze your transactions with different restaurants to gain insight into how they make the buying process easier or more appealing, thereby attracting and retaining customers.
4. An old adage goes, "It takes a lifetime to establish trust and only an instant to lose it." Basically, it takes only one bad incident to undermine years of building a positive transactional relationship. Think of a transactional relationship where this adage holds true for you. How did you respond to the supplier after the incident (personal, professional, or customer)? Can you remember any extraordinary efforts made by a supplier to meet your needs? Have you remained loyal?

5. Technological progress in communications over the past 15 years has dramatically reduced the cost of global phone and Internet communications. How would this drop in communication cost affect the make-or-buy decision for goods? For services? Create a list of occupations that could be affected by this dramatic, ongoing techonomic trend.
6. Techonomic metrics allow you to combine the utility of the technology with the cost to deliver it at different points in time. This allows you to develop a performance trend line. Trend lines are important for evaluating market potential, as there are typically limits to the costs the market will bear due to competition from direct or substitutional products. Techonomic trends often resemble an "S" curve: a region with slow progress, followed by a region of intense development, ending with a region of market maturity and slowing growth. Many failures can be attributed to entering the market too soon, with a technology that did not meet consumer expectations — or too late, after a competitor had become well positioned in the market. Name a product or market where you have observed the "S" curve effect. Substantiate your observations.
(Example: The delivery of bandwidth to U.S. residences has experienced the "S" curve over the past 15 years. Modems, computers, and infrastructure have combined to create a performance improvement of many orders of magnitude, but the pace of infrastructure build-out is now slowing, since everything from text to audio to images to video can now be delivered to the home in real time. Throughout the expansion, the economic component — monthly service cost — did not vary widely, while the bandwidth of service increased several hundred fold.)
7. Develop your own techonomic metric to measure the most economical transportation mode for a long trip. Compare air travel to automobile travel or any other mode of transportation you might want to use. The units of the comparison metric should be miles/$. This metric can encompass the transportation costs only, or you could add in the costs of your time in your metric to gain insight into why alternatives are chosen beyond the first order of price.

REFERENCES

1. Liberty Fund, Ronald H. Coase, Biography, in *The Concise Encyclopedia of Economics,* The Library of Economics and Liberty, Liberty Fund, Inc., 2002, http://www.econlib.org/library/Enc/bios/Coase.html.
2. Werin, L., The bank of Sweden, in Economic sciences in memory of Alfred Nobel 1991, Presentation speech, The Nobel Foundation, 2005, http://www.nobelprize.org/economics/laureates/1991/presentation-speech.html.
3. Saxe, J.G., The Blind Men and the Elephant, Wordfocus.com, 2005, http://www.wordfocus.com/word-act-blindmen.html.
4. Moore, G.A., *Crossing the Chasm,* Rev. ed., HarperBusiness, New York, 1999.

Section 2

A Techonomic Perspective of History

3 Organizational Evolution Resulting from Technological Advancement: A Timeline

INTRODUCTION

If I have seen further it is by standing on the shoulders of Giants.

—Sir Isaac Newton

In 1986, I started a small technology company called TeleRobotics International, Inc. I had recently left a national laboratory to start on a journey with an unknown destination. In hindsight, it is amazing to consider how many of today's common business technologies we did not have then.

We did not have a cell phone. We did not have a fax machine. We did not have a Web site. We did not have a local area network. We did not use e-mail. We did not have an inkjet printer. We did not have a CD or CD reader, burner, or player, much less a DVD. We did not have a color computer display because we did not have a color computer. Overseas telephone calls were measured in dollars per minute, not cents per minute. We did have a hard drive — bigger than a breadbox and holding all of 10 megabytes! And we invested in a used laser printer costing more than the rest of our office equipment and computers combined.

It was not that we were too small and too poor to obtain these things. Well, actually we were too small and poor, but the main reason we did not have those devices is because they were not widely available. All of those commerce-supporting technologies were either invented, perfected for mass use, or cost reduced for wide distribution in the past 20 years. *From the perspective of eternity, 20 years is a very short time. From the perspective of business, 20 years is an eternity.* Reread the list of "did not haves" and think about your daily business endeavors. Your organization would be noncompetitive if all these devices were removed from your office. In fact, you might be behind if only one of them was removed from your business today!

In this chapter, some of the greatest technology breakthroughs in history are categorized and sequenced. For clarity, the technology timeline will be presented in a graphic form revealing how many of these innovations rest upon preceding progress. Historical technology developments are grouped into four major classifications: energy, communications, computation, and community. These four classifications represent the fundamental pillars upon which organizations build. Rather than being randomly selected, these four organizational pillars extend from individual characteristics first described as the "four-square" principle by Danforth, discussed later in this chapter. We evoke the biology-to-business metaphor again, extrapolating the four-square principle from the single individual (organism) to the organization (collection of individuals) to establish the foundations of the four-square principle for organizations.

These four organizational categories provide the focus for the creation of techonomic metrics in the following chapter (Chapter 4). We shall consider how measurable advances in these key endeavors changed the way society was organized to pursue life. This look back in time is neither comprehensive nor definitive, but it is provided to illuminate the application of techonomic metrics as a way of understanding organizational change.

A TIMELINE OF TECHNOLOGY

Figure 3.1 shows a timeline of significant technical advancements.[1-13] I created this technology timeline by listing the most significant technologies upon which our society is based. Once the list was generated, Internet searches were used to determine the approximate date when these important innovations became popularized. During the process of the search, more key developments were added to the list, and a few were removed, being considered of less importance. ***Though the inclusion or exclusion of any technology in the timeline is justifiably arguable, the ease with which the Internet provides access to this information is undeniable.*** After the first compilation of the timeline list was complete, the grouping of developments into four categories provided a framework for further understanding: energy, communications, computation, and community. The timeline reveals these categorical building blocks are linked to each other, giving rise to waves of organizational progress.

Three conclusions emerged as the line stretched out:

1. Technology advanced slowly between 2500 BC and 1500 AD. Early community developments at the dawn of recorded history and slow but steady advancements in communications marked this time period.
2. Many technology advances were stimulated by military demands.
3. Following the introduction of the printing press, economical written communications rapidly increased the rate of technological innovation in all four pillars related to organizational evolution.

The direct and easy access to the information needed to create a timeline of this type should encourage you to think in broader ways related to your endeavors. Create your own "technology timeline" for an endeavor of interest to you so that you can

Organizational Evolution Resulting from Technological Advancement 33

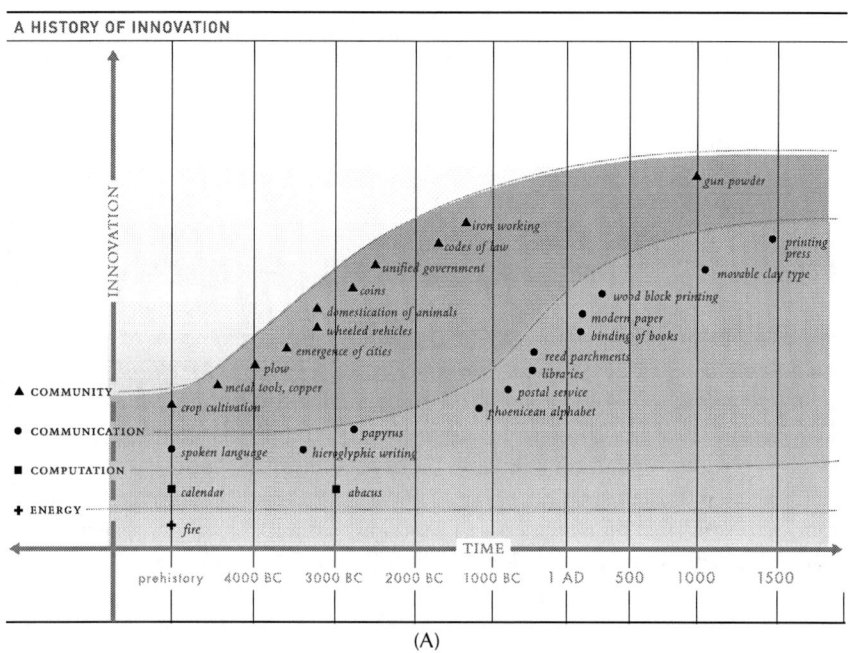

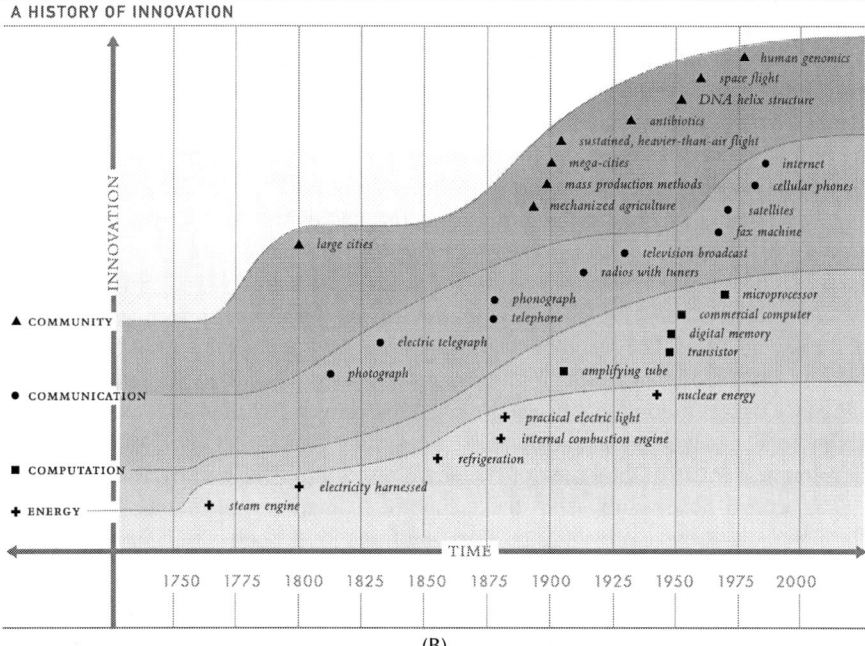

FIGURE 3.1 (A) A history of technological breakthroughs from prehistory to 1500 AD within the four-square principals. (B) A history of technological breakthroughs from 1750 to present within the four-square principals.

see how your field has developed. The process will broaden your understanding of the developments that support your efforts, and it will also reveal how the pace of development is increasing in every field. This access to information is necessary to develop trustworthy techonomic metrics — without reliable data, the metrics are simply guesses.

Consider what the timeline reveals about the progress of technology throughout history. A 4000-year period of limited technological development between 2500 BC and 1500 AD begs the obvious question: why? If we return to the fundamental assumptions of techonomics (see Chapter 1), the requirement for increasing information is prominent. Technological advance is rooted in information, as one development builds on previous foundations, standing on the shoulders of giants. *This period of 4000 years, 200 human generations, of very limited technology advance can be attributed to limited communications, hence limited growth in and availability of information.* The key development that completely changed the capture and dissemination of information was the invention of the movable-type printing press (credited to Johann Gutenberg, 1455). As a result, information could be duplicated for a fraction of the cost of laborious scribal copying on paper or other media (clay tablets, animal skins, etc.).

Clearly this new technology, the movable-type printing press, was more than a technological advance; it was an economic one also. The Bible was the first document Gutenberg printed. In Old Testament days, the Torah (the first five books of the Bible) was hand printed on a material called vellum. Vellum was made from stretched animal skins. *To create a scroll of the Torah required the skins of about 50 sheep and is estimated to have cost $300,000 (in today's money).* The invention of paper had reduced the material cost centuries ago, but the labor costs for hand copying remained high. The movable-type printing press reduced labor costs per printed page by orders of magnitude overnight. Not only was technology advancing, but also the process of discovery and invention was becoming widely recorded.

As humankind grew in its ability to record and distribute information by more than word of mouth or laborious inscribing, the amount of available information (and technological knowledge) began to grow. What has followed in the half millennium since Gutenberg has been a continuous acceleration of information creation and dissemination. The recent arrival of the Internet has caused another cost reduction on orders of magnitude for information access and distribution. Hence, the influence of technology on our economy and society is stronger than ever before as technological development itself accelerates. This increasing pace of technology development leads to the "Law of Accelerating Returns" coined by Ray Kurzweil.

A second observation from the technology timeline was that military efforts through the ages have proven to be a driving force in technology advancement. This observation reinforces the concept of the military as a leading indicator of technology development described in Chapter 1, but the timeline provides a different perspective of the same conclusion. It has been said that history is the story of man's inhumanity to man, and war is certainly the arena for much of that inhumanity. But our history is complex; technology developed through the demands of the military has also been used to save and enrich life. In many ways, military endeavors reflect the history of man's interaction with man. We learn from the interaction of cultures, even if it is on the battlefield. Long ago, devoid of many books that could disseminate information, the world's people commonly interacted through trade and military action. Having superior weapons and protection was the key to a society's survival. Whether society was organized by small clans of extended families or was a city-state of considerable size, defense was top priority. It remains so today.

The reality is, people have always learned from each other on the battlefield, and then some dimension of that knowledge has passed into wider application. If one culture developed a harder metal, its weapons became superior, and others copied it. If one culture developed gunpowder, it had a distinct advantage over its enemies, and the enemies adapted. If one country developed atomic weapons, it was able to subdue others for a time, and then others developed them also. Even today, many technological developments related to materials, space exploration, flight, electronics, biochemistry, and others too numerous to mention begin as military research and later filter to the consumer.

A society typically places high economic value on protection. The cost associated with the creation of an "ultimate weapon" is not bounded by the realities of the commercial marketplace (refer back to this concept in Figure 1.5). As a result, the entire economic well-being of a nation can hinge on pursuit of military protection. Money becomes "no object"; becoming the first to develop a capability is the supreme goal. The investment in technology paid off for the U.S. when the bomb developed by the Manhattan Project forced Japan's surrender and saved thousands of lives that would have been lost in a land invasion. The economic inability to make competitive military technology investments in the Cold War resulted in the fall of the Soviet Union without military engagement.

Private sector businesses must operate within financial limits that do not constrain militaries. Product prices must be realistically balanced by what the customer will bear. Enter the continuous advance of technology and the economic power of mass production. What was one-of-a-kind, expensive, and experimental in the recent past for a military application becomes widespread, affordable, and routine today. Moving forward a few years, the same technology may become ubiquitous, cheap, and indispensable. Microprocessors are a great, recent example of this techonomic trend. Developed for aerospace military systems, they migrated into top-end commercial applications and today are in everything from microwaves to watches.

A broad example is space flight and all its associated technologies. The space race was first entered as a response to foreign military threat. Our nation put men on the moon to prove we could do it! While travel in space has not yet become economically feasible for the masses (and probably will not until energy requirements are radically reduced), many developments related to spaceflight have migrated to mass application. Satellite weather monitoring, satellite telephone communications, satellite television networks, microelectronics, Velcro, global positioning systems, freeze drying, and insulating materials are just a very few of the hundreds of developments resulting from the pursuit of space travel — stimulated by the desire for military supremacy.

The third observation from the timeline was that all four pillars supporting organizational evolution advanced rapidly once the mechanism for communication provided by the printing press was available. The timeline reveals a first wave of community development in the shadows of prehistory, with key developments like cultivation, language, and rule of law. Over the next 4000 years, incremental improvements in communication technology lead to the printing press breakthrough (papyrus, paper, ink block printing, movable type). Once the power of the written word reached the masses, all four pillars expanded rapidly and significantly. First energy (steam power, internal combustion, electricity, atomic power) and then, on the platform of electronic calculation, the pillar of computation was built (analog tubes, transistors, microprocessors, computers). Electricity and ubiquitous computing allowed the reinvention of mass communications, first with broadcast media (radio, film, television) and then with interactive mass communications (Internet). The form, composition, and structure of organizations have evolved through each of these waves. Before we analyze the techonomic trends numerically for each of these four organizational pillars, we will establish the rationale for the selection of these principles as the keys to organizational evolution.

THE FOUR-SQUARE PRINCIPLE: ORGANISMS/INDIVIDUALS

In the classic book *I Dare You* by William H. Danforth (1931), the keys to fulfilled individual living are represented by a four-square checkerboard.[14] ***Danforth encouraged a philosophy of personal success based on balancing four key areas of living: physical (the body), mental (the mind), social (relationships), and spiritual (the soul).*** The square was a clear representation of equal-sided balance that Danforth encouraged. Figure 3.2 shows the four-square principle for individuals. This symbol was used as the logo for his company, Ralston Purina, and it can still be seen today on their animal feed products. Over 70-years later, *I Dare You* is now in its 36th edition, giving credence to Danforth's philosophy of personal balance.

An organization (or society) is a collection of human individuals. Continuing the analogy between biology and business, organisms and organizations, we will extend Danforth's four-square model from individual to organization. We classify technology development into four key endeavors that have supported the development of organized society throughout history.

THE FOUR-SQUARE PRINCIPLE: ORGANIZATIONS/SOCIETY

Figure 3.3 displays the four-square principle for organizations, extrapolating from the principles for individuals shown in the center. ***Extending Danforth's four-square principle for individuals, the four attributes to combine for successful organizations are energy, computation, communication, and community.***

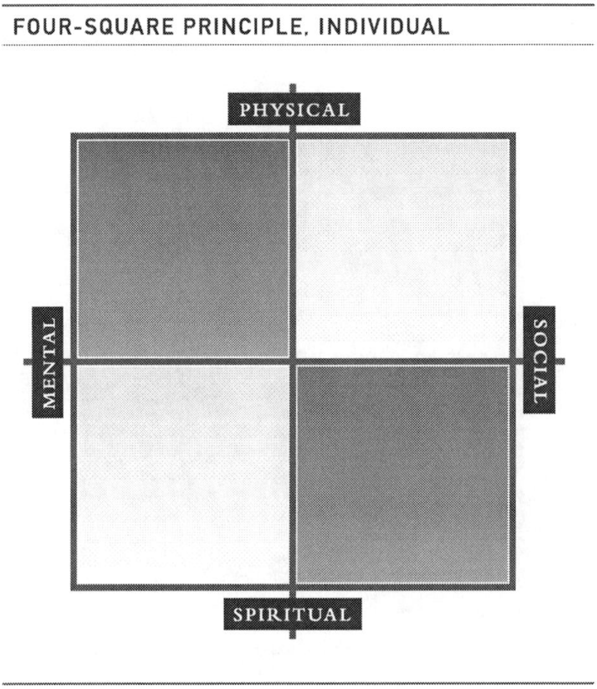

FOUR-SQUARE PRINCIPLE, INDIVIDUAL

FIGURE 3.2 The four-square principal for individuals.

Physical — Energy

The organizational analogy of the individual's physical strength is energy in various forms. ***In modern economic terms, energy is prosperity.*** Where markets and governments have allowed new energy sources and conversion methods to be deployed, standards of living have progressed, and prosperity has resulted. Where energy technology has not been embraced due to poverty, ignorance, corruption, or over-regulation, living standards have lagged. In the world of physical labor, energy determines our productivity. A road can be built by hand at the cost of thousands of man-hours, or it can be created in a day with the use of modern machines powered by energy. The same principle holds for planting/harvesting crops, constructing housing, making clothing and furniture, etc.

Energy sources can be classified into two broad categories: stationary (central, fixed energy source) and mobile (typically used for transportation). The distinction is important, given the rise in mobility and the increasing amount of energy consumption dedicated to transportation. It is also another example of how new human endeavors are built upon the availability of new technologies. Without internal combustion engines, heavier-than-air flight (airplanes) would not have been possible, nor would automobiles have penetrated into the culture. As we develop techonomic metrics to review society's energy progress, we will look at both stationary and mobile sources.

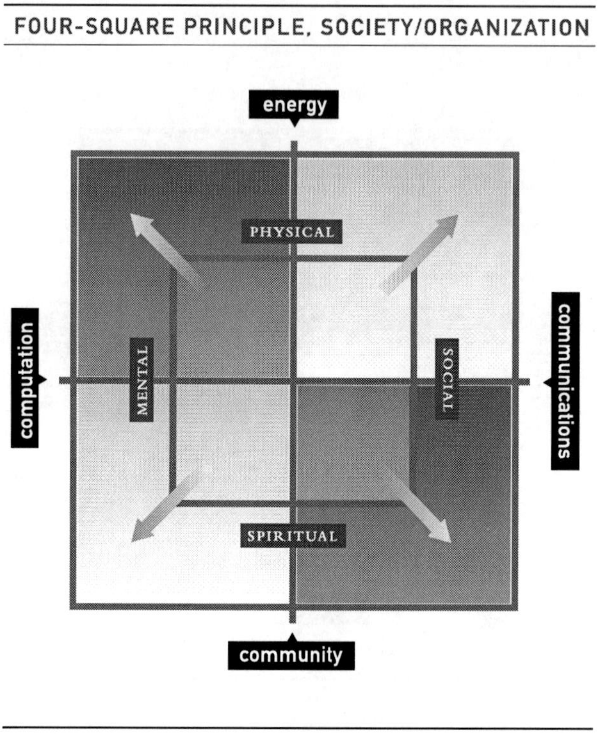

FOUR-SQUARE PRINCIPLE, SOCIETY/ORGANIZATION

FIGURE 3.3 The four-square principal for organizations.

The advent of efficient electric power distribution has allowed central power generation to serve the needs of the masses for their stationary energy requirements (mainly home and industrial needs). These central generating facilities can obtain their fuel from many sources, but switching fuel type is very expensive if not impossible. Techonomic metrics that combine the key elements of total cycle cost provide insight into our energy past and future options as a means to compare alternative fuel sources.

The harnessing of energy ultimately displaces individual physical labor, enabling society to be more productive and apply its human resources to other endeavors. The progress of technology reveals a cause-and-effect relationship between the harnessing of new forms of energy and the resulting change in human activities, often with a displacement of workers into new endeavors. From sails displacing oarsmen to steam-powered locomotives replacing horse-drawn wagon caravans, each major infusing of a new power source/conversion method has changed human life. Today, we observe the dichotomy of a world of energy haves and have-nots. Energy use in much of the third world has not changed throughout the ages: wood and dung for cooking and heat, animal power for some cultivation. Without the release from physical labor afforded by embracing energy technologies, time is limited for mental pursuits.

Mental — Computation

The individual attribute of the mind is mirrored by the computational capability of society. Surveying history, we see numerous developments in the understanding and implementation of mathematics. From numbers to represent counting, to geometry to track heavenly bodies, to the abacus, the electronic calculator, and the computer, mankind has made progress in leveraging mental capabilities to pursue human endeavors. In very recent history, the electronic computer has become a lever for the mind in the same way that harnessing steam power, internal combustion, and electricity became levers for the physical body. *The ramifications of mental leverage far surpasses the implications of physical leverage.*

While physical leverage helps us subdue our environment, thus freeing up time for other pursuits, mental leverage provides us the ability to work together in new ways. Mental leverage enables us to discover new horizons never before possible (consider advances in recombinant DNA research, particle physics, and astrophysics). We can now create machines and processes that actually are used to create the next, more advanced generation of machines and processes. *Technology has become the partner of imagination to reshape the world.* Richard Sennett, in his book *The Culture of the New Capitalism,* states the implications of this advance in technology capability directly: "By the 1990s, thanks to microprocessing advances in electronics, the old dream/nightmare of automation began to become a reality in both manual and bureaucratic labor: at last it would be cheaper to invest in machines than to pay people to work."[15]

In his book *The Age of Spiritual Machines,* Ray Kurzweil projects the computational capabilities that might be expected if electronic developments continue to advance at a rate comparable to the last 40 years.[16] He anticipates that a $1000 personal computer in the year 2020 will have enough memory and computational capability to store and take action on all the life experiences of a single individual (sights, sounds, thoughts, etc.). He further projects that, if current trends continue, that a $1000 personal computer offered in the year 2040 will be able to store and take action on all the life experience of all people living on Earth at that time! Even if Kurzweil is wrong in his exact timing or magnitude, the trends are evident, and the implications are staggering. In some ways, they are already coming to pass through the power of one system (the Internet) to access all systems.

Social — Communication

The individual attribute that Danforth calls the social side corresponds to communication from the technology perspective of the organization. Communications provide the foundation of interaction that sustains cooperation in an organization. Organizational communication has evolved in many ways as technology has brought about change through new possibilities. Through the ages, most enterprises were small because they required personal interaction to sustain focus and provide direction. This began to change in the nineteenth and twentieth centuries as technology provided new ways to communicate: newspaper, telegraph, telephone, radio, and others. Now organizations had methods to communicate across distances without great time delay, and new organizational principles could be enacted.

Technology continued to advance, now at greater rates because larger numbers of people could collaborate in its progress. Organizations could now expand their boundaries, no longer bound to word-of-mouth communications in a single location. In the early days of my career, common management practice was to place project teams in close physical proximity to maximize the interchange of ideas and strengthen cooperation. This was true when the most efficient form of communication was direct conversation: a stop by the office or a face-to-face meeting. With the advent of e-mail, voice over the Internet, teleconferencing, and video teleconferencing, the cost of communicating and monitoring remote operations has so diminished as to revamp management practices. With no financial communications barrier to outsourcing efforts, organizations are tending to globalize their operations at an increasing rate, seeking cost-effective labor wherever they find it. Survival of the fittest organization, in a competitive world, compels organizations to find the best match between quality and cost to meet their production and labor needs. Cheap and instantaneous communications have eliminated the global barriers to entry for mental labor and for many forms of manufacturing labor.

SPIRITUAL — COMMUNITY

While advances in technology can certainly be tracked for energy, computation, and communication, what is the spiritual correlation for an organization? Community. Extension of the individual characteristic of spirituality to the broader organization may not seem as evident as the other analogs, but it is even more important. Individuals grouped by purpose, mission, location, or nationality become a definable community. The unity of these individuals determines the vitality of the organization. Although the most cohesive communities are still composed of people who come into physical proximity to one another, community in the broader sense is not restricted to this. It is defined by purpose and shared experience as much as proximity.

It is not always easy to determine the unity of vision, purpose, and direction within an organization. My observations conclude that strong, healthy organizations exhibit a unity of purpose that attracts committed members to the community. Whether the organization is a city, company, people group, religious institution, or nation, the thriving entity exhibits a community characterized by harmony, vitality, and growth.

The most obvious and readily obtainable measurement of organizational community is size — population, number of employees, number of members, etc. A church may monitor the health of its community by the number of members, while a business may monitor its performance based on revenues or profits. In any case, there are a small number of key parameters that will indicate if the community is vibrant. For the organization to grow, it must be able to attract and sustain its members, attract and retain customers, attract and retain citizens, procreate and extend an ethnic base, etc. This measurement makes no judgment on the "goodness" of the organization with the general observation that organizations involved in worthy pursuits are the same ones that stand the test of time.

The history of technology advance is a story of developments that allow traditional communities (i.e., cities) of ever-increasing size to come together and sustain life. Through the ages, technology has advanced in sanitation, agriculture, medicine, transportation, energy production, communications, etc. to support exponentially increasing population. In most cases, before the population increased, the technology to sustain it emerged. When population outpaced the technology, innovations in disease prevention or agricultural methods emerged to support the next generation's expansion. Are these trends sustainable? Will technology provide answers via more efficient utilization of resources as it has in endless cycles in the past? Will economic realities of limited resources cause nations to limit procreation? Or will the historical pattern of increasing human population be changed by unforeseen circumstances — natural or manmade?

A look back in time reveals that the growth of communities has followed the emergence of the technologies providing their sustenance. Small villages emerged after the cultivation of crops and the domestication of animals; larger cities (> 500,000) emerged with the aid of steam power; megacities (> 1,000,000) rose from the harnessing of electricity; and today, the entire population of the world is in reach of satellite communications. Perhaps you will use techonomic insight to understand current population trends and shape the future.

In the next chapter, the process of developing techonomic measurements will be introduced and developed for the four fundamental organizational pillars: energy, computation, communication, and community. These techonomic metrics allow trends to be mathematically observed, and they reveal patterns of interaction where the extension of one organizational pillar rests on the progress of another.

SUMMARY

TECHNOLOGY TIMELINE

The technology timeline groups major advances by energy, computation, communication, community, and military. The phasing of these developments partially reveals how they are built upon each other and push each other forward in a continuously advancing spiral. Communication advances lead to energy discoveries that lead to community advancement that results in more energy discoveries/needs that lead to computational development that results in communication advances. Military research appears to be a key impetus for the cycle, since it occurs outside the typical constraints of competitive economic pressure. It is driven more intensely by competition to defend/conquer.

THE FOUR-SQUARE PRINCIPLE FOR INDIVIDUALS

This representation of personal balance adapted from Danforth's *I Dare You* describes the four fundamental qualities, and activities, of a healthy individual. Physical, mental, social, and spiritual pursuits, in a wholesome balance, yield productive and satisfied individuals.

The Four-Square Principle for Organizations

This representation of organizational balance is extrapolated from *I Dare You* as a basis for understanding proper balance in the culture of the organization. Energy, computation, communication, and community development, in balance, lead to a healthy and productive organization.

Key Terms

Four-Square Principle

Individuals	Organizations
physical	energy
mental	computation
social	communications
spiritual	community

QUESTIONS

1. The list of technology "did not haves" early in this chapter raises the question, "How could business be conducted effectively without these aids?" Describe the impact on your business endeavors of the past decade's diminishing cost of voice communication.
2. On the personal side, make a list of childhood "did not haves" and compare them to things you now have that make life easier for you (automatic faucets, automatic flush toilets, microwave ovens, etc.).
3. Figure 3.1 is a timeline of major technology developments that have changed society. Create your own timeline in a field of endeavor that interests you (use the Internet, personal observations, reference materials, etc.). What does the timeline say about the pace of change in your field? What key developments are on the horizon that will change how your field of endeavor is performed or is organized?
4. Large cities did not arise worldwide until after energy was harnessed to produce and transport food in quantities great enough to sustain these large communities. The migration from rural to urban areas continues today, and interdependence between urban dwellers is also increasing. What are the greatest demands society is placing upon technology in order to allow greater urban populations to flourish? What are the greatest vulnerabilities of urban civilization?
5. Military developments have been the consumer market's leading indicator of products to come. What products do you use today that trace their initial development/implementation to military funding?
6. Create a four-square personal assessment by drawing a four-sided figure where the length of each side represents your current efforts at physical, mental, social, and spiritual development. Balancing the four-square is

the goal. Create your perfect square and sign it as a personal goal. Compare the two squares. You have before you the foundation of a great personal development plan.
7. The four-square principle can be extrapolated from individual to organization. Draw a four-square assessment (refer to Question 6) for (a) the current organizational state of the nation, (b) the current organizational state of your company or a local group you work closely with. From your assessment, where would you focus your resources to balance/improve these organizations?

REFERENCES

1. Miller, D.R., World history timelines, Din Timelines, Miller Internet Publishing, 2004, http://www.din-timelines.com/.
2. Bellis, M., Inventors, About, About Inc., 2005, http://inventors.about.com.
3. Davies, R., Paper money, Origins of money and of banking, University of Exeter, 2005, http://www.ex.ac.uk/~RDavies/arian/origins.html#paper.
4. Grohol, J.M., History of Discovery, Psych Central, 2005, http://psychcentral.com/psypsych/Circulatory_system#history_of_discovery.
5. Bulaevsky, J., Pendulum clocks, The history of clocks, Arcytech, 2003, http://arcytech.org/java/clock_history.html.
6. Rogers Refrigeration, Refrigeration history, http://rogersrefrig.com/history/html, 2005.
7. Ament, P., 19th Century innovation timeline, The Great Idea Finder, 2005, http://www.ideafinder.com/history/timeline/the1800s.htm.
8. Department of Molecular Biology, History of antibiotics, Princeton University, 2001, http://www.molbio.princeton.edu/courses/mb427/2001/projects/02/antibiotics.htm.
9. Science Museum, Part III — Convincing the world, First Flight, Science Museum, 2005, http://www.sciencemuseum.org.uk/on-line/flights/first/after.asp.
10. Best, S., Brown, Z., and Lindsay, R., Reactors: Modern-Day Alchemy, Argonne National Laboratory Web Site, Department of Energy, http://www.anl.gov/Science_and_Technology/HistoryAnniversary_Frontiers/alchemy.html.
11. May, P., DNA, School of Chemistry, University of Bristol, 2005, http://www.chm.bris.ac.uk/motm/dna/dnac.htm.
12. Swenson, L.S., Jr., Grimwood, J.M., and Alexander, C.C., Project Mercury, National Aeronautics and Space Administration, 1989, http://spaceflight.nasa.gov/history/index.html.
13. Lander, E.S. et al., Initial sequencing and analysis of the human genome, The Human Genome, *Nature,* 409, 860–921, MacMillan Publishers, 15 February, 2001, web reference http://www.nature.com/nature/journal/v409/n6822/full/409860a0.html.
14. Danforth, W.H., *I Dare You,* 36th ed., American Youth Foundation, St. Louis, MO, 1991.
15. Sennett, R., *The Culture of the New Capitalism,* Yale University Press, New Haven, CT, 2006.
16. Kurzweil, R., *The Age of Spiritual Machines,* Penguin Group, New York, 2000.

4 Creating Techonomic Metrics

INTRODUCTION

A techonomic metric (abbreviated as TM) is a data measurement that provides insight into trends by combining the impact of technology advance on the economics of a given endeavor. The TM concept and a general process for developing TMs were introduced in Chapter 2. This chapter extends your understanding of this process by constructing TMs for the four historical pillars of organizations revealed in the preceding technology timeline. The TM is a means of comparison for an endeavor in one of several ways:

1. Trends for an endeavor over time: A TM can monitor the same endeavor at different points in time, revealing the impact of technology and economics. For example, how much did it cost to distribute information via a book at different periods in history? Techonomic metrics reveal the telling example of the ancient $300,000 Torah, today reduced to the $0.85 Gideon Bible.
2. Different technical approaches to the same endeavor: A TM can provide measurements of two competing approaches to the same endeavor, for example, how much it costs today to distribute information via a traditional book, or CD-ROM, or online.
3. Different cultural approaches to the same endeavor: A TM can provide measurements for performing the same endeavor in different cultures based on infrastructure, labor costs, transportation costs, communication costs, etc. For example, how much does it cost to distribute information to most of a society's population via the infrastructure available?

A good techonomic metric includes components of technological performance and economic cost for all key elements of the endeavor. To be actionable, a TM should be objective, and the data upon which it is constructed should be verifiable. When historic data is used that precedes contemporary currency (i.e., dollars prior to ~1930), the economic component of the TM should be based on a definable economic quantity, such as labor hours, land utilization, or individual human capacity. A useful TM will reveal a trend for a complex endeavor in simple, compelling terms. It may also clearly reveal where a technology breakthrough will move an endeavor from the experimental realm to mass acceptance because of cost reductions or performance enhancement.

THE TECHONOMIC SWEETSPOT

It is not sufficient for a new technology to simply work. It must work at a price the market will embrace. Great new technology takes the market in one of two directions. Either it opens a new field of endeavor (like space exploration), or it makes common practice endeavors that were previously accessible only to the elite (such as telegraph for distance messaging). Sometimes, great technology simultaneously opens new fields and rapidly becomes economically accessible to the masses (like books from the printing press). A great technology is so compelling that, as we look back in history, its introduction and massive application appears to coincide. What distinguishes a great technology from the merely good one? Significant increase in benefit vs. cost over other existing approaches is the indicator of a breakthrough technology.

Figure 4.1 is a two-dimensional representation of performance benefits (technology improvement) vs. value proposition (economic cost reduction), showing lines of constant techonomic metrics (performance/cost). As the techonomic metric increases, the potential for market acceptance of the endeavor also increases. While there are many good ideas and good opportunities, the great ones lie in the region where both performance and value proposition are significant — and coincide. When we survey opportunities for resource deployment, this region of greatest opportunity is the area to seek.

In the formative years of my company, we developed a video camera system that performed image pan and tilt without any moving parts. This device used electronics to select and correct the portion of interest from a wide-angle fisheye lens. Our first commercial deployment of the system was targeted at video security applications. The system required $3700 of electronics to perform the image processing for a standard video signal (NTSC 640 ×480 pixel resolution image quality). The product had two shortcomings in its first embodiment: it cost too much relative to competing mechanical systems, and its resolution was not sufficient to see clear details from segments of the images. While these items are obviously important, we had conquered hundreds of other challenges just to get the product to the marketplace. But the marketplace could not care less about what obstacles we had overcome. The marketplace cared only about performance at a competitive price. In its purest form, the market is a rational, unemotional filter for performance and cost.

After a good introductory quarter for the product (the distribution pipeline filled), we had a dismal second quarter, and it was evident that we had to refine our offering or cease to exist. Timing coincided with the earliest popularization of the Internet. A company called Netscape had just provided the first massively utilized browser, and personal computers were connecting to the Web everywhere. In adapting our product to this new environment (the Internet), we recognized the opportunity to provide a similar function for mass distribution over the Web at a much-reduced cost to the consumer. By using other people's electronics (personal computers rather than our own electronics) and distributing software over the emerging network (the Internet), the capability to pan, tilt, and magnify a 360° image was provided at a fraction of the original hardware cost.

Creating Techonomic Metrics 47

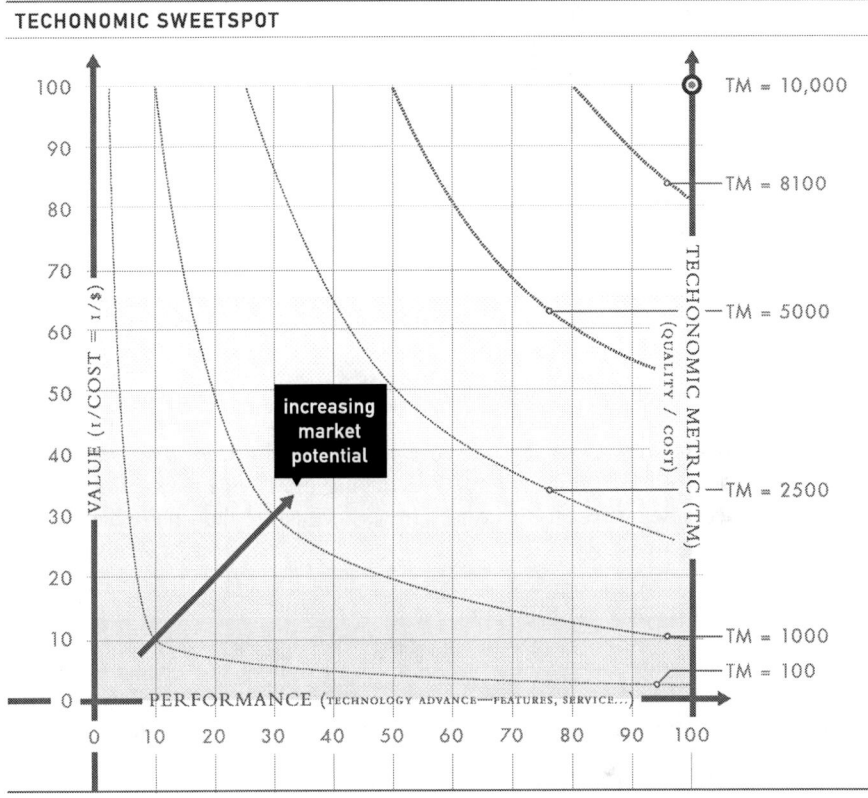

FIGURE 4.1 Cost vs. performance showing lines of constant techonomic metric measure.

In the first hardware-based system, old image transmission standards for television limited the resolution of the captured image and resulted in a limited-quality display for the end user. Transforming from the old standard (NTSC video signals) to the emerging distribution network for digital images on the Internet, much higher resolution input images were possible, improving the quality of the result. Using scanned photographs rather than video, we were able to increase the resolution 4 to 8 times. Technology provided multiple order-of-magnitude improvements in both cost and performance, moving the product much closer to the techonomic market sweetspot. Using software rather than hardware, we were able to reduce delivery cost to the end user from $3700 hardware to a few pennies per download, while simultaneously improving the resolution 4 to 8 times. Revisiting the performance/cost relationship, these combined developments improved the performance proposition 6 to 7 orders of magnitude and eliminated the need for closed-circuit system operation. This tremendous shift in the performance/cost relationship for the experience delivered to the customer was the key to refocusing our resources. The product was later named iPIX and continues to be widely used for interactively viewing 360° images of homes, automobiles, apartments, and resorts on the Internet.

AN EXAMPLE: DIGITAL PHOTOGRAPHY TECHONOMIC METRIC

I had a deep interest in digital photography, since the transition from film to digital was the key to ease of use for mass producing the iPIX 360° image. Without economical digital cameras, the iPIX process required labor-intensive film scanning to produce an immersive image. In the early 1990s, few digital cameras would satisfy our resolution needs, and none were in mass production. We required a two-megapixel minimum image, a camera under $1000 for mass distribution, and a special lens to capture a wide field of view (fisheye) that was not typical of anything on the market. We lacked resources to develop the whole camera, but we did have the resources to develop the lens when the "right" cameras became available. Moore's Law (the Law of Computational Ubiquity; see Chapter 5) was progressing, so it was only a matter of time before the cameras would be on the market. But when? Specifically, when would we need a fisheye lens for our process that would allow digital production of 360° images, greatly reducing the labor content for the iPIX process? An accurate answer to this question was critical to the timely deployment of our limited resources.

DIGITAL CAMERA TECHONOMIC METRIC

Many factors determine the performance of a digital camera: size, battery life, ease of use, sensitivity, color balance, resolution, software support, etc. The simple TM developed to follow the advance of the digital camera used the following assumptions:

1. Purpose was to determine when commercially viable digital cameras would emerge.
2. Fundamental economic component is the $ per image element provided.

$$TM_{\text{Digital Cameras}} = \text{Pixels/Cost [pixels/\$]}$$

where
- TM: techonomic metric
- Pixels: number of pixels (picture elements) in camera imager
- Cost: retail price of the digital camera

This TM served to predict the timing of lens development needed to coincide with the availability of satisfactory digital cameras. The input data was easily accessible on the specification sheet for every new product introduced. In hindsight, it also revealed a very interesting "S-curve" for digital camera performance. (An "S-curve" is two-dimensional diagram which displays a slow increase until a certain catalyst causes rapid change to a new performance level, hence resembling an "S".) This **insight from the marketplace was the seed from which the Theory of Techonomics germinated.** The threshold TM to provide a satisfactory 360° image capture system at a reasonable consumer price was two million pixels in a camera

Creating Techonomic Metrics   49

priced at less than $1000 (TM = 2000 pixels/$). When this search began in the early 1990s, such a large TM seemed outlandish and inconceivable. Due to broader technology trends discussed in Chapter 5, this threshold was rapidly cleared. The historical progression included:

- Mid-1970s: Kodak develops solid-state sensors (small pixel counts, large cost; 32 pixels/$3000 = 0.01 pixels/$).
- 1987: Kodak releases first profession digital camera products (1,300,000 pixels/$10,000+ = 130 pixels/$).
- 1994: Apple releases first consumer digital camera product (640 × 480 NTSC television resolution, 300,000 pixels/$1000 = 300 pixels/$ with imager based on repurposed video camera imager).
- 1996: Numerous providers of consumer megapixel cameras (1,300,000 pixels/$1000 = 1300 pixels/$ with imager based on electronics specifically for digital photography).
- 1997: iPIX consumer threshold TM of 2000 pixels/$ surpassed.
- 1999: Price reductions via competition and mass manufacture (2,000,000 pixels/$500 = 4000 pixels/$).
- 2005: Price point stabilized and resolution satisfies printing needs (4,000,000 pixels/$500 = 8000 pixels/$).

Figure 4.2 plots the TM for digital cameras as a function of time. It shows the rapid advance in price performance of these cameras that coincides with the introduction of these products to the masses. Such an introduction results in cost reductions associated with mass manufacture and also accelerated innovation as competitors differentiate their new products. The curve also reveals that the window of rapid performance advancement is not necessarily sustainable. *At a certain point, customers are satisfied with the general capability of the product, so technological innovation gives way to market positioning.*

The marketplace penetration of new technical products tracks the techonomic metric with a time lag. Once the performance/cost metric satisfies the desires of the market, rapid adoption occurs. There lies the value of the techonomic metric as a predictor of future economic value. The digital camera market and its aggressive undermining of the film camera position is an excellent example. Figure 4.3 reveals the past decade's camera sales, by year, for film and digital cameras.[1]

Obviously, the sea change from film to digital has now occurred, lagging the techonomic metric by 2 to 3 years. Given the total market for cameras, one would expect a leveling off of annual sales as the techonomic metric reaches stability and the market reaches saturation. The period of rapid adoption is followed by a period of slow growth in demand.

This example is provided to introduce the process of creating a viable techonomic metric. In review:

1. Survey the technology contributors to the endeavor you wish to monitor.
2. Determine the one or two most important, broad contributors to the endeavor.

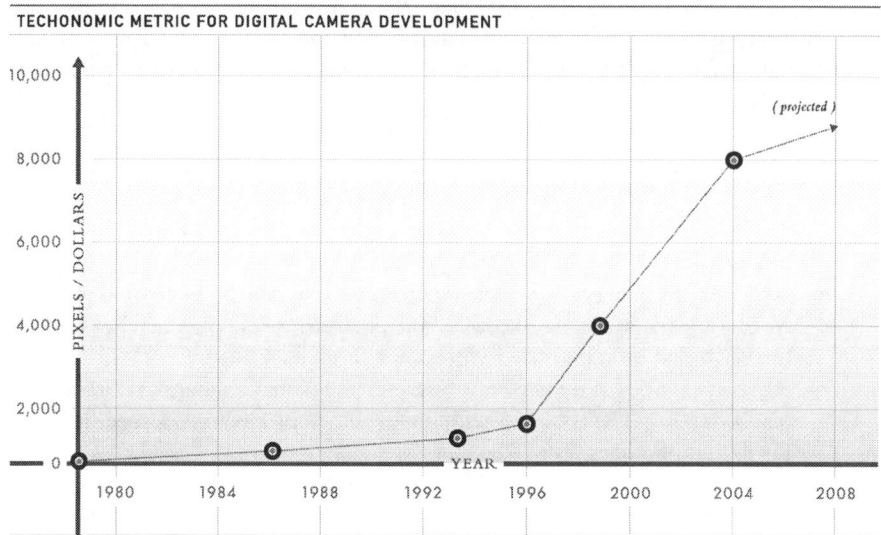

FIGURE 4.2 A techonomic metric for digital camera development.

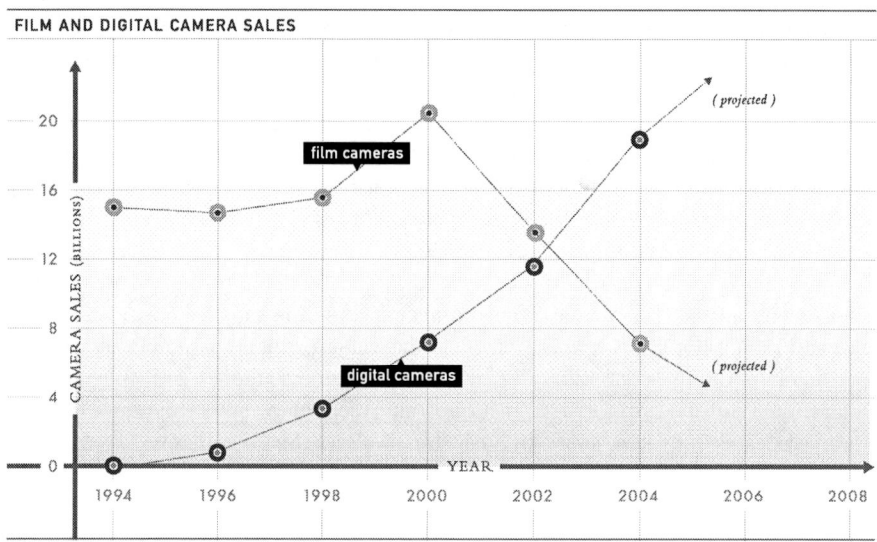

FIGURE 4.3 Camera sales by year for film and digital cameras over the last 10 years.

3. Seek an economic measurement of the contributors.
4. Make sure you can get objective data that allows you to construct TM comparisons between time periods, competitors, or cultures.

Note that economic measurements can be quantified in terms of currency or human labor (output per person or per hour). But output alone does not constitute a techonomic metric for an endeavor. Output related to *cost or effort to perform the endeavor* does. In surveying techonomics throughout history, metrics in terms of human labor are more useful than metrics based on currency that no longer exists or has no comparative value.

MILITARY: TECHNOLOGY ADVANCE WITHOUT ECONOMIC CONSTRAINT

As we have discussed, military supremacy has always been a high-priority societal goal, because it means protection and economic advantage. Military technology performance has been well documented throughout history and provides a wealth of data to support the development of a military TM.

Here is one obvious military performance metric: the destructive power of common weaponry through the ages, calculated in terms of kinetic energy. This metric does not consider any economic basis, only the power associated with a single armament. As such, this is not a techonomic metric but an interesting measure of the advancement of military technology. Comparing the bow and arrow to the largest tested nuclear weapon shows that the kinetic energy content has increased sixteen orders of magnitude.[2]

- Hand-thrown projectile: 100 ft-pounds = 0.0000505051 HP-hrs = ~$10^{(-16)}$ kilotons (c. prehistory)
- Cannon charge: 5 to 10 pounds gunpowder — ~0.000003 kiloton equivalent (c. 1000 to 1500)
- 21,000 pound massive ordnance air burst conventional bomb: (power classified, 0.0105 kilotons if all TNT) (c. 2000)
- Nuclear weapons: 450 kilotons (c. 2000)

While destructive power alone reveals the key historical technology infusions (gunpowder, TNT, nuclear), it does not indicate other important aspects of warfare (delivery of armament, command and control, etc.). To develop a techonomic metric for military endeavors, one must focus on the purpose of military action. Military action is often taken to subdue occupants and control territory. The key economic component of the TM is the soldier (assuming wars are typically won by attrition at some level).

MILITARY TM: HISTORIC

Assumptions:

1. Purpose of the endeavor is to control territory.
2. Fundamental economic component is the soldier.

$TM_{Military}$ = territorial span of control/individual soldier [square meters/person]

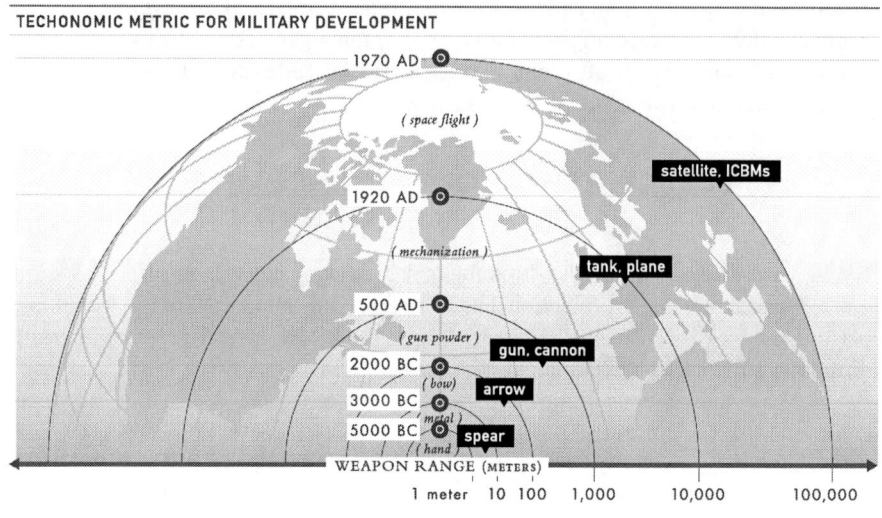

FIGURE 4.4 The military spans of influence for the individual soldier over time revealing the impact of technology.

Other metrics could be developed to trace the impact of technology on the battlefield: army size, battles per year, casualties per year, etc. Let us just consider a military TM that combines measurable endeavor results (area of territory controlled) with the economic unit providing the results (individual soldier). Certainly, in the history of warfare, it is not the individual, but the magnification of the individual by technology that determines the distance over which an individual can project control. *Major shifts in the TM result from introducing new technologies to the performance of the endeavor.*

Figure 4.4 shows the military span of influence for the individual soldier over time. This figure approximates the control projected by a soldier and the major technologies that impacted reach, measured by the area coverage of the armament.

- Hand combat (c. 5000 BC) — 1 meter^2/soldier (weapon range ~1 meter)
- Hand combat plus swords and hand-thrown spears (c. 3000 BC) — 100 m^2/soldier (weapon range ~ 10 m)
  - Technology — metal for weapons and protection
- Hand combat plus crossbows and arrows (c. 2000 BC) — 10,000 m^2/soldier (weapon range ~100 m)
  - Technology — invention of the sling, bow, etc.
- Fortified facilities and catapults (c. 1000 BC) — 100,000 m^2/soldier (weapon range ~300 m)
  - Technology — fortifications, armaments and new strategies
- Explosive weaponry (guns, rifles, cannons, bombs) (c. 500 to 1500 AD) — 1,000,000 m^2/soldier (weapon range ~1000 m)
  - Technology — gun powder

- Mechanized warfare (c. 1900 to 1940 AD) — 100,000,000 m²/soldier (weapon range ~10000 m)
  - Technology — internal combustion propulsion (tanks, troop transport — WWI)
  - Technology — heavier-than-air flight (WWII)
- Automated warfare (c. 1970 AD) — 1,000,000,000,000 m²/soldier (weapon range ~1,000,000 m)
  - Technology — electronic automation and guidance
- Space warfare (c. 2000 AD) — 100,000,000,000,000 m²/soldier (weapon range ~10,000,000 m)
  - Technology — space flight, laser armaments, satellite surveillance, power storage and conversion

This military TM emphasizes the projection of might for a soldier, considering the individual soldier as the fundamental economic unit in a military conflict. With each major advance in military technology, the coverage of the individual soldier was magnified. Over the ages, advancing technology has increased the soldier's reach by 7 orders of magnitude and coverage by 14 orders of magnitude.

Note the time interval between arrivals of paradigm-shifting technologies has decreased significantly through the ages. From the beginning of recorded history to the introduction of gunpowder (~4000 BC to ~1000 AD) was a period of about 5000 years. The mechanized augmentation of warfare occurred in the 1900s (~900 years later). The atomic age emerged in the 1950s (~50 years later). The space age of warfare — with its high-tech communications, control and weapons delivery — matured in the 1970s (~25 years later).

In the broad scope of time, techonomic metrics measure the repetitive process of discontinuous advance caused by the implementation of new technologies. The series of "S-curves" that result start slowly (like warfare based on muscle-power) and exhibit incremental improvement through many years, until a new technology supplants the old way of doing things (like gunpowder). The new technology yields a rapid performance increase, until saturation of the new approach leads to another period of incremental improvement (the top-flattening of the S-curve). The S-curve cycle may begin again with another innovation (mechanized warfare or atomic bombs), but at some point these improved performance benefits also saturate until their effect is supplanted by another significant technological innovation. These multiple discontinuities are not evident in the broad-brush resolution of a chart that covers all of time, but they would be evident in performance advances localized in time around any on of the major technology introductions.

Unfortunately, these generations of ever-more-effective military technologies leave in their wake a legacy of ever-more-powerful capabilities. These legacy technologies are surpassed by national militaries, but remain highly destructive in the hands of small organizations set on terrorizing others. *The march of technology provides commonplace devices in the hands of the masses that were not even available to the most sophisticated military organizations two generations ago (Internet information, cell telephones, video surveillance, certain explosives,*

***radio-controlled vehicular toys*, etc.).** Hence, technology advance can simultaneously provide creature comfort to the masses and be a source of terror in the hands of the disenfranchised.

Even if military technology does not seem to have limits, its economic justification typically will. When we observe the current international proliferation of atomic weapons, it seems evident that all want to join the club, but hopefully, none want to be the first to deploy these weapons. So the motivation to build ever-more-powerful weapons is minimized, and the resource focus moves to weapons miniaturization, material surveillance, and intelligence. In other words, there is always a point where enough is enough for a given technology deployment, and the cost/benefit for incremental technology extension is not justified — even for the military.

We move our attention from the leading techonomic indicator, the military, to the development of techonomic metrics for the other pillar endeavors fundamental to the evolution of organizations (refer to previous the chapter, Figure 3.3). Techonomic advances in energy, communications, and computation have provided the framework for ever-expanding communities. Techonomic metrics for these endeavors provide an understanding of the trends of history, validating the techonomic analysis process and also provide a number of "how to" examples for developing techonomic metrics in order that you might learn to apply the methodology.

ENERGY: SIDE 1 OF THE ORGANIZATIONAL SQUARE

"No energy is more expensive than no energy."

— **Dr. Homi Bhabha**

As our review of military history pointed out, for the better part of 5,000 years, animate labor provided most of the energy for human endeavors. With the minor exceptions of wood burning for heat and cooking, waterwheels for grain grinding, windmills for water pumping, and isolated uses of coal and natural gas, the world ran on animate labor (humans and animals).

Energy development can be summarized by major technological impacts: discovery of fire (prehistory), development of external combustion engines (typified by the steam engine, 1750 to 1770), development of internal combustion engines (typified by the automobile engine, 1870 to 1890), commercialization and mass distribution of electricity (1910 to 1930), and harnessing of nuclear power (1940 to 1960). Each of these major developments changed both the living patterns and the fundamental fuel sources/requirements of society.

The invention and application of the steam engine (1765) increased energy consumption dramatically, providing physical leverage far beyond animate labor. Early in the twentieth century, two additional energy developments, electricity and internal combustion engines, allowed large quantities of energy to be distributed to the populace and provided transportation for the masses. These technologies increased fuel consumption for predominantly hydrocarbon-based systems. The most recent key development in energy is nuclear electric power generation, providing

Creating Techonomic Metrics

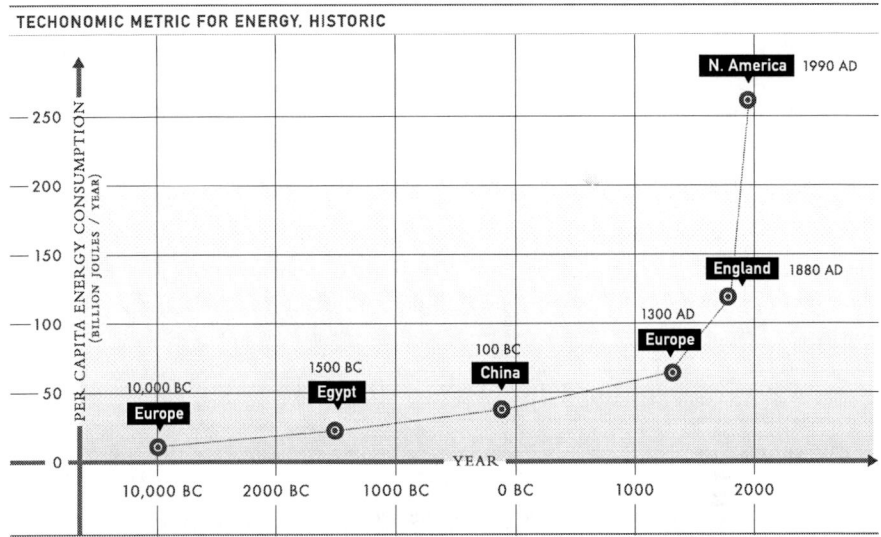

FIGURE 4.5 Typical annual per capita energy consumption throughout history.

even greater energy density and generating capacities to meet the needs of growing communities. An energy techonomic metric for an historical perspective tracks per capita annual energy consumption through the years. The greatest increases in per capita consumption have occurred as new technologies made it possible to consume greater amounts in the pursuit of improved standards of living. The TM for energy increases coincident with the spread of steam power, the deployment of internal combustion engines and the availability of electricity. Figure 4.5 shows the typical annual per capita energy consumption during different stages of history.[3]

ENERGY TM: HISTORIC

Assumptions:

1. Purpose is to track energy usage in human endeavors.
2. Fundamental economic component is the individual consumer.

$TM_{Energy\ Historic}$ = annual energy consumption / person [billion joules/year per person]

Harnessing energy to magnify human effort brought many related changes to society. Transportation, construction, agriculture, military endeavors, and ultimately computation were transformed by the availability of copious and new distribution forms of energy. As portions of the populace were freed from physical labor, mental endeavors increased — which in turn resulted in further technologic advances.

Techonomic metrics for comparing current energy production methods must consider many elements, including:

- Fuel cost (unit energy costs for extraction, transport, refinement)

- Conversion cost (unit energy costs for plant capital and operations costs to convert fuel from chemical/nuclear form into mechanical or electrical energy)
- Environmental/regulatory cost (unit energy costs for environmental protection of air, water, or containment of waste products)
- Mobility factor (suitability for stationary or mobile applications)

The sum of the fuel cost, the conversion cost, and the environmental cost per unit of delivered energy provides the comparative techonomic metric for energy. The mobility factor is a "go" or "no go" determination of fitness for an intended purpose.

ENERGY TM: COMPARATIVE

Assumptions:

1. Purpose is to compare energy sources for human endeavors.
2. Fundamental economic component is the $ per unit energy provided.

$$TM_{Energy\ Comparative} = \text{delivered energy per unit cost [Megajoules/\$]}$$

We can find information for delivered fuel cost from numerous sources, but conversion costs vary widely based on application requirements.[4] The smaller/simpler the application, the less our fuel selection decisions are affected by conversion cost. For instance, conversion capital requirements and operating efficiencies for personal home heating vs. central electric power generation are substantially different. Given fuel prices and conversion unit efficiencies for home heating systems, we can easily compare techonomic metrics based on the annual cost of heating a home.[5]

These metrics cover the major financial elements of fuel and conversion, but they do not consider environmental effects. In an unregulated system with no environmental constraints, considering only the price of energy required for the desired service finishes the evaluation. However, in most contexts, we must certainly consider environmental factors. For a central power generation facility, environmental and operational regulations are playing an increasingly prominent role in fuel selection. ***Unknown costs of regulation (nuclear) and changing environmental requirements (coal) have shifted fuel selection for new generating capacity in the U.S. over the last two decades to different sources (natural gas).*** The Three Mile Island accident (1979) and the Chernobyl accident (1986) halted orders for new construction of nuclear power plants in the U.S. A nuclear-powered electric generating facility has not been built in the U.S. in the last 20 years. The economic uncertainty of regulatory, safety, and licensing expenses has lead to the demise of the domestic, commercial nuclear industry.

However, this has not been the case worldwide. France, for example, now generates over 75% of its electricity from nuclear facilities with a standardized design that reduces uncertainties. In 2000, there was as much electricity produced from nuclear energy as was produced from all energy sources worldwide in 1961 (2438 billion kilowatt-hours.)[6]

Similar environmental challenges, resulting in economic and regulatory uncertainty, are also evident in the use of coal for electricity generation. Particulates, sulfur dioxide, nitrous oxides, and greenhouse gasses are all contributors to air pollution that are being regulated to protect the environment. The emission targets are not stationary and are becoming increasingly tight as environmental effects become better understood and environmental lobbies gain political power. Over the past decade, the U.S. has shifted toward natural gas for electric generation because it provides the least risk in the environmental/economic equation — even though it is more suitable for many other, more demanding (mobile and chemical production) applications.

Here is a clear example of the techonomic metric yielding an answer in a nontraditional manner, by revealing the contributing areas that cannot be numerically evaluated. This is RISK. *Whenever there is a strong contributing element to a techonomic metric that cannot be reasonably estimated, the risk to making a resource deployment is considerable.* If a government determines that it wants to create an environment of uncertainty for use of a technology like nuclear power generation, or *any* technology, then that industry will not advance in that country. Likewise, if a government creates an environment that clarifies and supports the application of a new technology — like France did for nuclear power generation — then the industry will advance quickly. France now has 58 nuclear reactors, and they provide about 77% of the country's electricity. In 1973, France was relying on traditional fossil fuels for over 80% of its electric energy needs.[7]

The forces of economics ultimately enter the global marketplace, as indicated by the words of the late Indian physicist Dr. Homi Bhabha, "No energy is more expensive than no energy."[6]

COMPUTATION: SIDE 2 OF THE ORGANIZATIONAL SQUARE

The second side of the organizational square is computation. For most of human history, computation has been done within the speed and capacity limits of the individual human brain. Computation acceleration beyond human input was not possible. While the knowledge base of the fundamentals of mathematics was advancing throughout the ages, means of providing computational leverage for the human mind lagged. For those interested in a detailed history of computation, Stephen White provides a most interesting account.[8]

The early abacus served as humankind's only computational lever for five millennia. It magnified human capacity by counting, carrying, and serving as a memory device. With the discovery of the logarithm, the development of the slide rule was made possible in the first quarter of the seventeenth century (William Oughtred, 1625). The next major advance was the mechanical calculator using gears and mechanisms to perform addition, subtraction, division, and multiplication (first mass produced by Charles de Colmar in 1820). Although human input was still required to initiate each calculation, some speed advance over unaided human computation was achieved.

Charles Babbage (1792 to 1871) provided a big step forward as he conceived the fundamentals of computing machines — machines that could be programmed to execute many computations without further human intervention. Babbage conceived numerous mechanical computers with innovations including memory, punch-card programming, and conditional jumps. His mechanical designs provided the conceptual foundation for the electronic computers that followed a century later.

Observing the computational speed of these methods throughout history, one realizes that the computational lever for the mind was very limited until systems were developed that used electricity as the basis of operation. One ancillary techonomic observation is that mechanical development of a process often precedes electronic development — the physical implementation precedes the virtual. In a way, things are understood in the tangible world before they are implemented in the electronic world. The invention of the vacuum tube by Lee DeForest (1906) was the gateway to electronic computing. Once an electrical representation for mathematical operations occurred, the speed of computation increased by orders of magnitude, and eventually the cost per computation decreased even more dramatically.

COMPUTATION TM: HISTORIC

Assumptions:

1. Purpose is to compare computational speed available for problem solving.
2. Fundamental economic component is time per mathematical computation.

$TM_{Computation\ Historic}$ = mathematical operations per second [ops]

One of the first digital computers, the ENIAC (Electronic Numerical Integrator And Computer) could perform 50,000 simple additions or subtractions per second.[9] The ENIAC was developed during World War II, representing another example of military demands driving rapid technological development due to absence of economic constraints. Early use of ENIAC included calculations related to projectile paths and supporting calculations for the hydrogen bomb. While ENIAC was unreliable due to the thousands of vacuum tubes required for it to operate, it ushered in the era of electronic computation, and the lever for the mind was unleashed. By the way, ever wonder where the word "debug" came from? To modern computer programmers, it means fixing a computer problem by correcting errors in computer code. It originated with the early, mammoth-sized, relay-driven computers. Moths would get into the circuitry and cause malfunctions, so computers had to be cleaned out to function properly: debugged.

The speed and miniaturization of computation has been on a rapid rise ever since the days of these early computers.

Physically ENIAC was a monster — it contained 17,468 vacuum tubes, 7,200 crystal diodes, 1,500 relays, 70,000 resistors, 10,000 capacitors and around 5 million hand-soldered joints. It weighed 30 short tons (27 t), was roughly 2.4 m by 0.9 m by 30 m, took up 167 m² and consumed 160 kw of power. As of 2004, a chip of silicon measuring 0.02 inches (0.5 mm) square holds the same capacity as the ENIAC.⁹

The development sequence and interactions are worthy of note. For eons, advances in computational capabilities were nonexistent, but the body of knowledge of mathematics and processes continued growing. The harnessing of electricity led to many discoveries and products, including the vacuum tube in the early twentieth century. Advances in vacuum tube and computer technology were slow at first; the boost came via a wartime effort requiring significant computation. The war prompted focused research and development in practical and applied electronic computation. Analog computers, requiring hardwired programming for each problem, appeared first at the turn of the twentieth century. In response to military needs, a new category of digital computers emerged, exemplified by ENIAC, and computational capabilities were increased by orders of magnitude.

Next, the slow, steady road to commercial applications began. Commercial requirements demanded higher reliability and reduced cost for general applications. At mid-century another military race began, the space race, and the need for miniaturized, light-weight, and reliable computing for space vehicle control fueled advancement. Related technological developments emerged (transistor and integrated circuit) directly addressing size (miniaturization), operating environment (reduced power consumption and heat load, increased component reliability), and programming flexibility (high-level software languages). With each related technology improvement, more commercial applications became economically viable. *In a nonintuitive twist, the Soviet Union had superior rockets reducing their need to limit the weight and size of the electronic control payload. As a result, there was less motivation to miniaturize electronic circuitry — a need that drove the U.S. program and subsequently resulted in the birth of the U.S. microelectronics industry.* These advances found their way into the commercial mass market a decade later with the introduction of the personal computer.

The journey of computational progress accelerated, as the technology was now in the hands of the masses, creating economic opportunities for innovations, both in hardware and software. With more commercially viable applications, economies of scale reduced hardware manufacturing costs dramatically, and a growing body of knowledge related to software solutions became available. The geometric expansion of worldwide computational capacity, initiated by the military-backed development of ENIAC, continues into the twenty-first century with little sign of slowing.

Figure 4.6 shows the computations per second available from a single device. Before the dawn of electronic computing in the 1940s, the speed of human input constrained the computational performance of the limited methods available. After the 1940s, the invention of the transistor and the integrated circuit created an ever-increasing performance metric for computation. Even in the absence of an economic component, one can easily observe the significant increase in computing power afforded by the emerging technologies.

A second metric, combining the impact of improving computational speed with the cost reductions that have accompanied these advances, provides a techonomic perspective on computational advance. While the speed of various technologies has

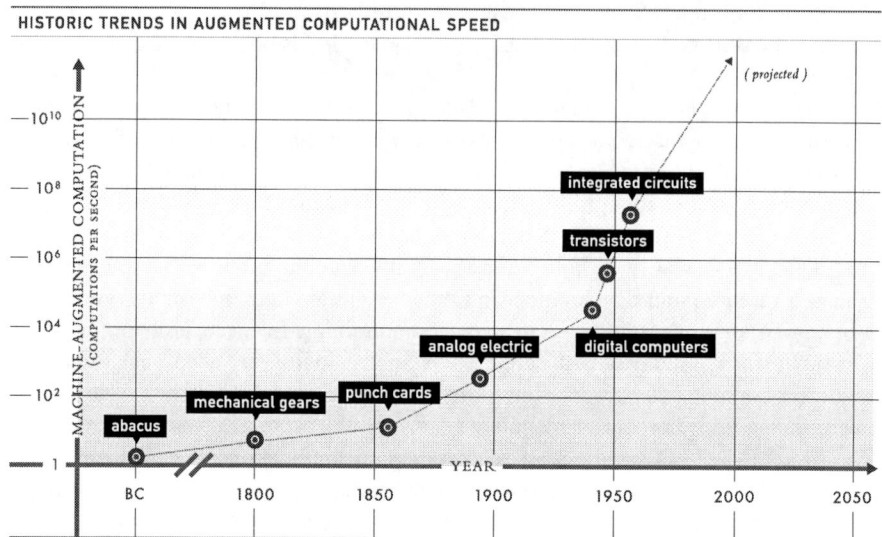

FIGURE 4.6 Historical computations per second available from a single device estimated from numerous sources.

increased exponentially over the past 50 years, implementation costs have simultaneously decreased. This has been due to the economics of implementation (aided by mass production), reduced material requirements, reduced energy requirements, and improved reliability. Combining the advance in technological performance with reduced implementation cost yields a complete techonomic metric of the advance of computational capabilities available for commercial applications. This metric is closely related to Moore's Law (see Chapter 5), which predicts that the cost of equivalent computing performance halves every 18 months.

Computation TM: Comparative

Assumptions:

1. Purpose is to compare computational speed per unit cost to track the techonomic progress of computation.
2. Fundamental economic component is mathematical computation speed of a device per dollar of cost for obtaining the device.

$$TM_{\text{Computation Comparative}} = \text{operations per second per dollar [ops/\$]}$$

The computational techonomic metric is a bellwether of the advance of digital "intelligence" in current and future commercial products. As long as this metric doubles every 12 to 24 months, as it has for the past 40 years, the anticipated shelf life of digital computational products will be limited.

The latest and greatest digital devices will be outdated in 3 years, replaced by something much more powerful and probably less expensive. This is one reason an increasing number of electronic firms have opted for contract manufacturing of their products. The production arrangements are more seasonal batches than permanent manufacturing lines. The effect of this trend is evident in consumer products ranging from personal computers to cell telephones, from personal digital assistants to flat screen televisions. ***Electronic product life cycles are about 2 to 3 years and shrinking. The importance of maintaining your position in the marketplace by "cannibalizing"*** (introducing your next product in the same marketplace, taking away sales from your own product to extend your market position) *your own leadership position* is of increasing economic importance. Intel realized this early in the 1970s and has followed this approach masterfully for over 30 years, regularly introducing new microprocessors to supplant their previous offering.

In 1993, Motorola was excelling with a leadership position in the rapidly growing cellular telephone industry. Their corporate experience with professional systems for two-way radio communications (police, fire, military) provided them the perfect technology base to aggressively distribute cellular phone systems for the public. Using this base, Motorola became a first mover when telephone deregulation occurred. Their revenues and profitability expanded rapidly for a few years during the initial cellular phone build out. But they based their approach primarily on the analog technology they had perfected over the years. They were reluctant to cannibalize their leadership position when digital transmission technology came on the scene. Nokia, much smaller and seeking an opportunity to impact the marketplace, embraced the digital transmission approach and rapidly gained market share. Thus, a company with far fewer resources and weaker market position turned the tables on a giant by embracing the technology trend that was inevitable. Motorola spent 5 years recovering and has now returned to a competitive position, but it must now contend with a formidable opponent limiting market share and profitability in a competitive industry.

In this same era, Kodak made a similar strategic decision. For decades, Kodak had maintained a leading position in photographic film, papers, and chemicals. The digital camera arrived in the photographic marketplace in the early to mid-1990s. Kodak's participation in these years was at best a "me too" approach, providing products that were a generation behind and focusing most of its marketing effort on a film system (the Kodak APS, Advanced Photographic System) that offered incrementally improved photographic quality. They also introduced the PhotoCD in an effort to capture the market from film to CD-ROM for digital information applications. All of Kodak's major thrusts at this time were intended to prolong a leadership position for photographic film while the marketplace was abandoning film as the capture-and-storage medium for images, replacing it with digital imagers and memory.

Kodak miscalculated the rapidity with which the quality of consumer digital photography would become comparable to film photography (resolution, storage cost, color fidelity — all changing at rates predicted by Moore's Law, see Chapter 5). Digital photography also heralded new possibilities for photography that were not even considered by those grasping to maintain the position of film (instant gratification, ease of use, keeping only the pictures desired, digital enhancement, etc.) In doing so, Kodak missed a significant opportunity to position itself as the processing and printing leader for digital images, thereby extending its paper and chemical ink businesses. Seeing the rise of digital photography, competition from many directions arose to provide personal photographic printing (Canon, Hewlett-Packard, Lexmark, others), digital image print processing (Fuji), and digital cameras (Nikon, Canon, others). In more recent years, Kodak has refocused its attention on the printing side of imaging and is regaining lost ground, but the nearly exclusive leadership position it held in film photography has now been supplanted by the digital age.

One company that seems to have learned many of these lessons is Apple Computer. For years they were recognized as a leading innovator of personal computer technologies, but they suffered long term in the marketplace because of their closed architectures and premium pricing. This led to less software support and reduced economies of scale. Apple has successfully introduced a line of MP3 players called the iPod. In the 3 years since its introduction, Apple has already revamped and expanded the product numerous times. They have provided a complete solution including a distribution Web site that combines the audio and video content from scores of record labels and thousands of artists. Their business model for success is based as much on media distribution as hardware sales, and according to their Web site, they have now exceeded 500,000,000 downloads of songs. Realize that the cost of goods for the electronic transfer is negligible except for royalties that are paid after the sale, and you begin to realize that Apple is creating a very profitable cash flow from its media sales in addition to its hardware sales. This is techonomic thinking at its best — creating and executing a viable economic model based on technology trends.

Long-term success is not guaranteed, since many competitors are entering this market, but Apple has established a strong leadership position by changing the way music is experienced (MP3 player vs. CD player) and changing the financial model for success by establishing a viable and rapidly growing distribution system that brings in significant residual revenues. They have an open system from the standpoint of the music media providers. They have an easily accessed and affordable system from the standpoint of the customer. And as electronic costs continue to drop due to trends and manufacturing scale, they are positioned to dominate a lucrative market for a season — 2 to 3 years. The flock of third-party, add-on accessories that have emerged due to the popularity of the iPod may position it to remain in a leadership position for a much longer time due to expanded capability, networks, and partnerships. Of course, there are other savvy companies out there. The forces of techonomic natural selection never cease in the free market.

These examples reveal the importance of understanding trends in the deployment of resources. ***Even the biggest industry leaders have to make choices on how to deploy their resources among competing demands.*** Such focus enables companies to repeatedly weather the storm of change resulting from the advance of technology and competitive threats. The focus *must be on the future,* not the past. Learn the lessons of the past, but apply them to where the technology trends are heading, not where the market has been. Those who wait for the market to lead find themselves running an exhausting race to catch up with ever-accelerating changes.

Summarizing the history of computation from a techonomic perspective yields the following interesting observations:

1. Although humankind increased in understanding of mathematics throughout the centuries leading up to the modern era, the mechanisms to exploit this knowledge were limited by slow, human input (abacus, geared calculators, etc.).
2. With the introduction of electricity into common use at the turn of the twentieth century, first analog computers and then more impressively digital computers began to leverage the human mind to address complex mathematical problems at high speed.
3. The technological developments replacing large, energy-intensive, and comparatively slow vacuum tubes with transistors and then microcircuits created an economical approach to computation, opening its possibilities to the masses.
4. The continued "shrinking act" of microelectronics, now in its fourth decade, is providing twice the computation performance every 2 years. This trend does not appear to be abating, and it affects the life cycle of every digital consumer product.
5. If current trends continue, by 2020 a desktop computer will have the storage and processing power of a human brain (storing every experience and able to make inferences on it at the rate of a human). Again, if trends continue to 2040, that same desktop computer will have the capacity of the entire human population.[10]

Just as the harnessing of steam lead to rapid growth in leverage for animate energy, likewise the harnessing of electronic computation lead to rapid growth in "computational" leverage for the mind power of humanity. The improvements in computation that society is utilizing could not have occurred without the harnessing of electricity (energy, side 1 of the organizational square). The knowledge that lead to the steam engine was captured and distributed by the printing press (communication, side 3). The printing press was the culmination of hundreds of small advances over thousands of years leading to paper, ink, movable type, etc. advancing communications. The pillars supporting organizational evolution also reinforce and accelerate each other.

We now rely on such computations to augment daily activities in thousands of ways. Machines are designed that are so complex they could not be created without the aid of existing machines. *One Digital Day: How the Microchip is Changing Our World* pictorially describes how digital computation is touching all aspects of life.[11] According to *One Digital Day,* the average American touches 70 microprocessors before lunch each day. ***If leverage of physical endeavors was humanity's greatest advance in the nineteenth century, augmentation of the human mind via electronic computation was the greatest advance of the twentieth.***

COMMUNICATIONS: SIDE 3 OF THE ORGANIZATIONAL SQUARE

The third side of the organizational square is communications. Communications parallels the social side of the personal square, since through communications we interact with others, have commerce with others, disseminate the body of human knowledge, and learn from others. The techonomic, organizational definition of communications refers in its broadest sense to the ability to capture information and disseminate it to others. The evolution of communications throughout history has been discontinuously accelerated by technology changes, the most important being the most recent one: the Internet.

Along with the proliferation of computing, the world is experiencing an exponential expansion of communications capability due to the expansion of the Internet. Only a few developments in the span of history can approach the impact of the Internet on communications. These developments include the first human speech, the first written languages, the first alphabet, and the invention of the printing press. In this review of communications history, we will emphasize the effect of the mechanically printed word upon society.

It is the mechanically printed word that led to publication of more history books than ever before, and that is what allows us to review history, including the history of communication! In fact, *before Johannes Gutenberg invented the printing press in 1438, there were only about 30,000 books of any kind throughout all Europe.*[12] Most of these were hand-copied Bibles or biblical commentaries. By 1500 (*less than 75 years later), there were more than 9 million books on a host of topics.*

Historians differ on the exact dates at which historical eras begin and end, but the end of the "dark ages" and the beginning of the Renaissance period can certainly be tied to the advent of the printing press and the information explosion it set off. More definitively, the press gave wings to Martin Luther's 95 Theses (1517), and the Protestant Reformation that influenced all of Western culture was under way. Books, information, and education were no longer available to only the few, but were economically within reach of the masses. There was great motivation to learn to read, and the ability to leverage the sharing of knowledge increased like never before.

But books did not originate with the printing press; they just became economically viable for mass distribution. Writing had progressed through many phases, including painted pictures in caves, stylized hieroglyphics carved on rocks, cuneiform impressions on clay tablets, ink on animal skin scrolls, ink on papyrus rolls, and ink on bound paper books. Each of these steps made writing a little easier, its duplication a little less expensive (in terms of human labor), and its dissemination a little wider. A techonomic metric for observing the historical progress of written communication before the electronic age could consider many of the following contributing elements:

1. Given the media and the language, the density of information per unit volume made possible by the combination of written elements
2. The approximate labor time required to duplicate a written message

3. The material costs of the writing elements and transport material
4. The skill level required to produce writing and to read the results

To create our techonomic metric for an historical review of written communications, we focus on the first two items from this list: information density per unit volume and duplication labor. Why? Item 3, material costs, is difficult to quantify in terms of today's dollars. Exotic materials (gold, silver, jewels, etc.) were not associated with common writing, although they were used for ornamental inscriptions. Item 4, the skill level of the duplicator, required literacy. At all times throughout history, the job of scribe required education, not just common labor.

Combining the information density as determined by characters per one square foot (per page), incorporating the thickness of the page as a function of the material and its binding, and then dividing this quantity by the time required to duplicate one square foot of information, the historic techonomic metric for written communications is obtained.

COMMUNICATION TM: HISTORIC

Assumptions:

1. Purpose is to estimate labor required for reproducing and distributing the written word in order to track technology's impact on the progress of written communication.
2. Fundamental economic component is human labor required to replicate a fixed quantity of written information using the medium available while incorporating the thickness of the media as an indicator of functionality.

$TM_{Computation\ Historic}$ = characters per unit volume/duplication labor hours [(characters/ft^3)/hour]

Figure 4.7 shows the techonomic metric for the history of communication, revealing trends in efficiency for duplicating the written word as a result of technology advancements in materials and writing techniques [(characters/ft^3)/hour]. The first advances are in information compression, moving from pictures to glyphs (representative symbols) to words for description. These early improvements in written communications aided in the common understanding of written language and the expansion of the labor force able to replicate it, but it did not impact the efficiency with which it could be produced. The next sequence of advancements improved permanence, as engraving on stone replaced cave paintings. Still, replication remained a slow and tedious, and it required skilled labor. Clay tablets addressed the engraving challenge by using imprints on a malleable surface and then baking the tablets for permanence. Clay tablets provided faster printing and were more portable than granite blocks, but there was still a lot of room for improvement. The invention of papyrus (reed paper, about 2900 BC) provided a semipermanent medium that could be written on with ink. The codex, a bound set of pages, arose somewhere in the second to fourth centuries AD. The codex allowed nonlinear access to written

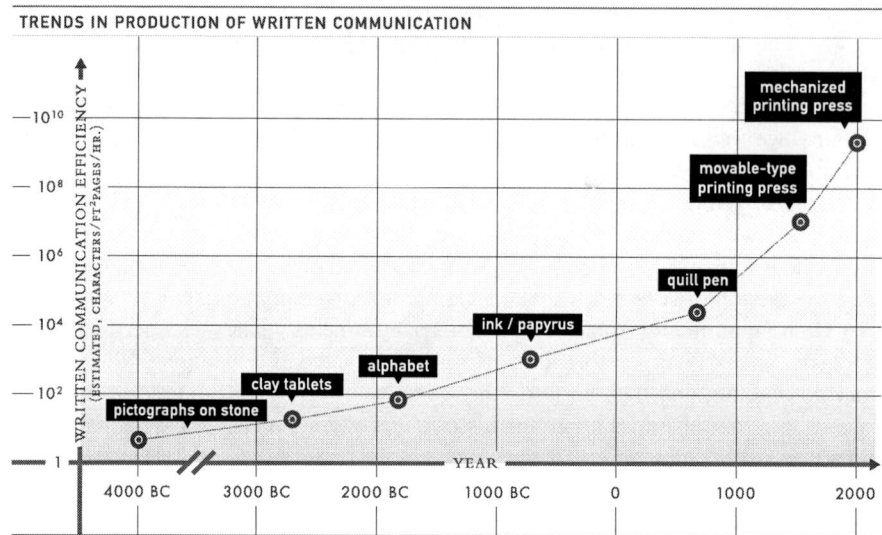

FIGURE 4.7 Historical communication techonomic metric revealing trends in the efficiency of duplication of the written word as a result of technology advancements in materials and writing techniques.

volumes and also was more easily stored and had a larger capacity. Vellum, a stretched animal skin, was also a widely used substrate for writing in the late Roman Empire (400 AD). This medium was often covered with wax and could be scraped and reused. Manufacturing improvements also made vellum more readily available than papyrus about this time.[13]

The rise of writing technologies was not smooth or uniform in all cultures and places. Paper was invented in China as early as 100 BC, but its journey to Europe and acceptance there was slow due to resistance from sheep and cattle landowners.[14] As late as 1221, a decree from the Holy Roman Emperor Frederic II declared all documents written on paper to be invalid because paper had been introduced to Europe by Moslem culture. But as techonomics would anticipate, the demand for paper caused by the printing press (1450) allowed the economic advantages of paper to overwhelm the cultural disfavor and the landowners' self-interests. The technology breakthrough for the duplication and distribution of the written word prior to the electronic age was the printing press.

Between the invention of the small, hand-worked printing press in the mid-fifteenth century and the bigger, more mechanized presses of the late nineteenth century, the distribution of written communication did increase. It was the twentieth-century advent of electricity and its associated information transmission developments that heralded the next major transformation in communications.

New ways of communicating emerged with the advent of electricity. The written word could be instantly transmitted over distances (telegraph, facsimile, Internet), and a copy machine could duplicate a page of information in a few seconds. The spoken word could be recorded and replayed, transmitted, transcribed, and broadcast worldwide. Still and moving images could be captured, reproduced, digitized, analyzed, and broadcast. The harnessing of electricity by computational equipment ultimately allowed all modes of communications to take a universal form — the digital bit — and to be disseminated on a universal, worldwide network: the Internet. Now words, music, voice, pictures, video, and information are encoded and reconstituted in a single, universal, digital stream that can be instantly delivered anywhere in the world.

The techonomic metric that provides insight into the magnitude and rapidity of this trend relates the bandwidth speed and availability to service cost for the individual. Possible components for the comparative metric for digital communications include:

1. Bandwidth available to consumer (kilobits per second, kbps)
2. Cost of service ($/month of service)
3. Percentage of population reached by the communication service (%)

Let us now analyze these possible TM contributors. The bandwidth available (item 1) determines the time to access a given amount of information. It determines the type of media that a consumer is willing to access (i.e., few will download video over a telephone modem, simply because of the time it requires). A number of competing digital communications systems — including telephone modem (56 kbps), ISDN (64 kbps), DirectPC satellite modem (400 kbps), B-ISDN (~1.5 Mbps), cable modem (~1.5 Mbps), and ADSL (~1.5 to 2 Mbps) — are vying for consumer loyalty and revenues. Currently, the downloading of live video streaming demands the greatest bandwidth for consumer distribution. Live video will fill the capacity of all available bandwidth distribution systems, but the faster ones (~1.5 Mbps systems) can maintain a continuous video stream without significant visible degradation.

Item 3, the coverage availability, is more a function of capital investment in infrastructure and consumer embrace than it is a function of technical merit. It may be important in determining how many customers there are for a given media form, but it is more an indicator of the status of the infrastructure than the capability of the technology, so it is not included in the comparative communication TM. The exception to this is the potential for a discontinuous technological breakthrough (i.e., ubiquitous wireless high-capacity bandwidth delivery, finding economical ways to deploy the last mile to the home of the fiber optic infrastructure, significant advances in compression methods, etc.). For this metric, coverage is presently constrained by wired infrastructure. But as we apply this metric, gazing into the future, we must keep in mind the possibility that a breakthrough wireless technology may be emerging. The bandwidth delivery speed and cost combine for the comparative techonomic metric for communications as described in the following paragraphs.

COMMUNICATION TM: COMPARATIVE

Assumptions:

1. Purpose is to compare digital data delivery capacity per unit cost for digital communication.
2. Fundamental techonomic components are the bandwidth delivered to the consumer and the monthly cost of the bandwidth.

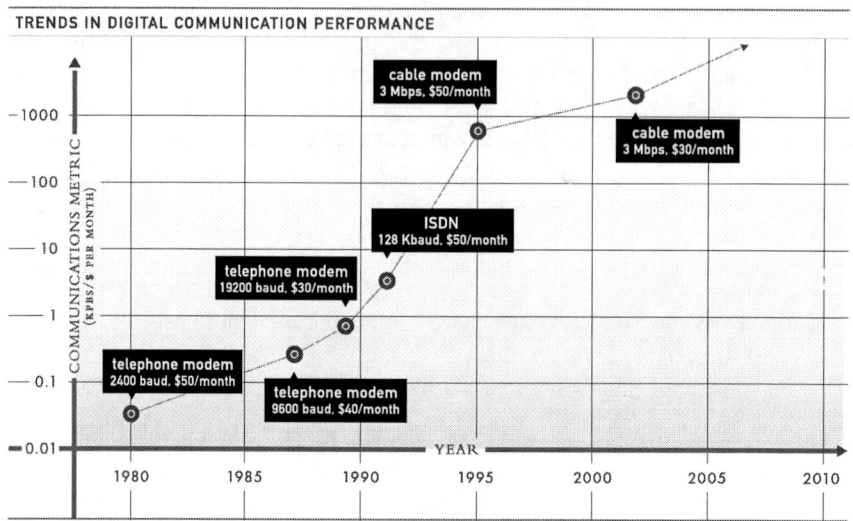

FIGURE 4.8 Comparative communication techonomic metric revealing recent communications cost reduction trends for consumers as a result of technology deployment.

$$TM_{\text{Communication Comparative}} = \text{bandwidth / cost per month [kbps/\$ per month]}$$

Figure 4.8 shows a comparative techonomic metric for communication, revealing recent communications cost reduction trends as a result of technology deployment [kbps/$ per month].

This monumental economic shift in communications cost resulting from technology advance has been a huge factor in accelerating globalization because of a considerable drop in communication cost. Once communication costs for verbal and data communications dropped significantly below the difference in local and international labor rates, the economic justification to relocate many jobs internationally was compelling. Barriers to labor that had existed for centuries due to communication costs in one form or another vanished in less than a decade. As a result, multinational companies followed the economics of lowest production costs. These costs are closely tied to direct labor costs and indirect labor costs (add-ons for vacation, retirement, healthcare, workman's compensation, unemployment insurance, etc.). Without the buffer of large international communications costs, it was not feasible to sustain the disparity in international wages for the same output. Only transportation costs associated with bulk or perishable items continue to play in favor of local production for manufactured goods in the U.S. economy.

It should come as no surprise then, why the U.S. is rapidly moving toward a service economy. Only the service sector (construction, healthcare, restaurant, education, delivery, retail, hospitality, government, etc.) requires direct personal contact that justifies the labor wage premium currently garnered in the U.S. Even many "service" sector activities that can be performed remotely are being transferred to international operations due to labor savings (telemarketing, technical support centers, accounting, drafting, etc.), since the communication cost barrier to entry has been virtually eliminated by techonomic advances.

Technology has created a leapfrog effect/opportunity for some of the underdeveloped regions of the world. The advent of wireless communications allows coverage of the urban areas in the developing worlds by telephone using modern systems. These wireless systems reduce the capital outlay required for hardwiring and eliminate the need for access to physical right of way.

It is also instructive to observe that the nature of communications itself underwent a monumental shift with (1) the introduction of electricity (resulting in telegraph, telephone, radio, television) and (2) the advent of digital representation and transmission of information (computers, networks, Internet, digital forms of all data types). These two modern communication technology waves, electrification and digitization, resulted in massive augmentation of the written word as the primary store of human knowledge. The fiber optic worldwide backbone also increased the communications density possible though a given cable, further reducing the cost of communications. Information changes resulting from advances beyond the printed page have been profound. Information has become:

1. Easier to generate
2. Disseminated via electrons (bits) rather than just atoms (paper)
3. Significantly cheaper to distribute
4. Easy to distribute to large audiences
5. Instantly transmittable around the world

We have come a long way from the 30,000 (or fewer) books possessed in the whole of Europe before the fifteenth-century invention of the printing press. As of 2002, the number of books available in the U.S. Library of Congress was estimated to be over 17 million volumes. Positing an average of 1 megabyte of text per book yields an estimate of 17 terabytes [10^{12} bytes] of written information in the Library of Congress. Perhaps even more indicative of the current explosion in production and dissemination of information is the estimate that the Internet passed over 440,000 terabytes of original emails in one year, 2003.[15]

Summarizing the history of communication from a techonomic perspective yields the following observations:

1. Until the printing press arrived in 1450, written communication via books was limited to the elite, and most human knowledge had to be transferred verbally and maintained by stories and tradition.
2. Printed books opened knowledge and education promoted literacy for the masses. Discoveries resulting from shared information contributed to the Renaissance, the "rebirth" of human creativity and intellectual advancement.
3. Electricity, and the myriad of communication devices it enabled, again expanded information distribution markedly.

4. Mass distribution of information, afforded by telephone, radio, and television, linked communities and regions more closely than ever before possible.
5. Massive digitization of information, accomplished by the availability of inexpensive computers, and the digital connection of the entire world, made possible by the Internet, eliminated communication barriers between nations for all forms of information. We are now a world community in the sense of communications.

The Internet supplants the traditional "rule of seven". (The experimental finding that you can reach anybody on Earth by following the introduction of the right acquaintance and their subsequent network through seven steps.) *The Internet is the network of networks, and the search engine is the ultimate "finder."*

COMMUNITY: SIDE 4 OF THE ORGANIZATIONAL SQUARE

In Danforth's four-sided individual square, the fourth principle is the spiritual. Every organization has a culture, a spirit so to speak, but it is difficult to quantify. Like good leadership, you know it when you see it, but it defies exact measurement. For this fourth side of techonomic analysis — the "spiritual dimension" of the organization — we will focus on growth of community. **Vibrant organizations, vibrant cultures, vibrant cities tend to grow** and flourish. Weak organizations, failing cultures, crumbling cities tend to shrink and diminish.

In prehistoric times, people lived as hunter/gatherers, foraging for food wherever necessary. With the advent of cultivated crops, people were able to settle and team together, and the extended family village arose. The village was limited in size (population) by the ability to grow and transport food. Further along in ancient history, the city-state was the community heart of human organization. Currency was invented, and trade increased. Folks banded together in cities for mutual protection, specialized in various tasks, and cities grew. But city size was limited by the people's ability to provide basic needs for all citizens. The land could provide only a certain amount of food within the limits of animate production (animal power). Perishable food products could be transported only to local markets. Without mechanized transportation, people could transport the large quantities of food needed to sustain large cities only by boat. As a result, the earliest large cities developed at the crossroads of river or ocean transportation.

The steam engine slowly removed this limitation, providing power to pump water for irrigation, operate mills for food processing, and drive locomotives to transport food over greater distances. Cities began to grow. Overall, societies remained largely agrarian, but the industrial revolution developed mechanized industry. Machinery began to transport goods, help produce crops, speed textile production, and shape the avenues of commerce: roads, canals, tunnels.

More people could be sustained in closer proximity, hence the emergence of large cities. Further specialization of tasks arose with increased commerce.

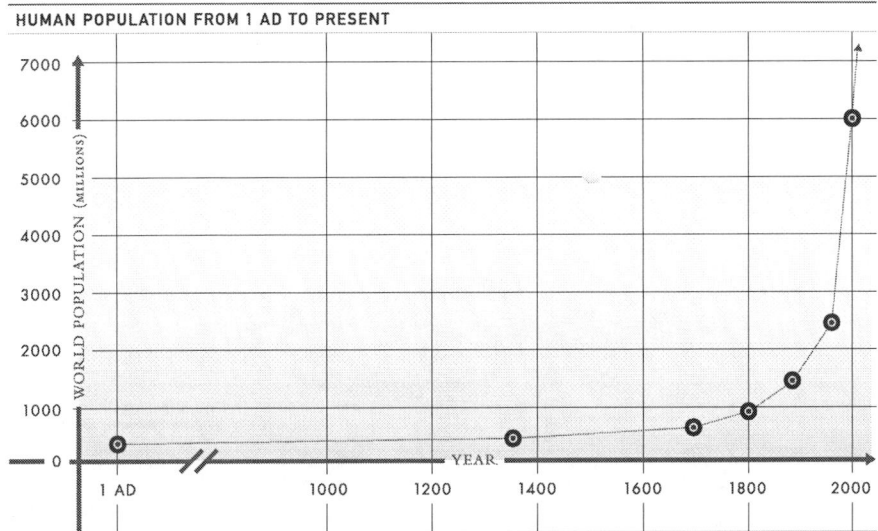

FIGURE 4.9 Historical world population from 1 AD to the present.

Community interdependence became a fact of life. The interdependence of family was now extended to the area inhabitants, as folks relied on each other for life's necessities. Shopkeepers relied on farmers for their food. Merchants relied on artisans for value-added goods. Artisans relied on farmers for raw material, and they relied on merchants to sell their products in the market. Farmers relied on shopkeepers for their clothes. While self-sufficiency was the norm in agrarian society, interdependence became the theme of urban life.

Figure 4.9 shows the historical world population from 1 A.D. to present.[16–17] This exponential growth has been accompanied by migration to urban areas. Figure 4.10 shows the number of cities with a population of over 1 million, over time.[18–19] The emergence of the world's first two cities with a population of over 1 million people (London and Beijing in about 1800 AD) coincides with the onset of the industrial revolution (ushered in by Watt's steam engine in 1765). By 1900, there were a dozen cities with a population of over a million, and by 2000 there were as many as 430 population centers with over a million residents. The world's largest urban center, Tokyo, is now estimated to have over 34 million residents.[20] Currently, in Earth's most populous urban areas, population density ranges between 25,000 and 100,000 people per square mile (Tokyo, New York, Hong Kong, Mexico City, others). The current worldwide population density is about 110 people per square mile.[21]

TECHONOMICS OF SUSTAINABILITY

Healthy growth is the organizational analog to Danforth's spiritual dimension for individuals. At the same time, there is a great deal of very justified concern and discussion worldwide about the sustainability of the growing population, given environmental impact and given the fixed quantity of resources available on Earth.

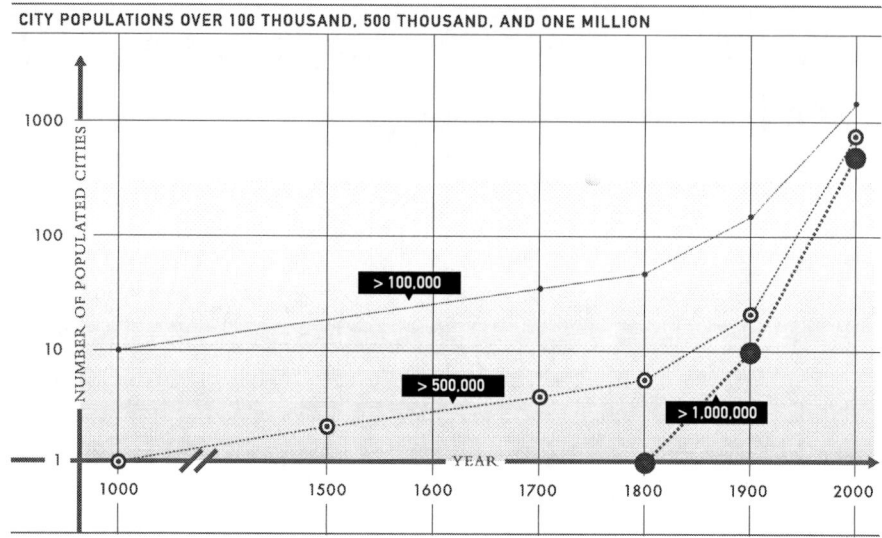

FIGURE 4.10 Number of large cities with a population of over 100,000, 500,000, and 1 million over time.

Big cities are mega-organizations — organizations of many organizations — but they are subject to the same techonomic laws that govern all organizations. These laws function in positive ways within the context of human freedoms: political, economic, intellectual, and personal.

This historic techonomic review reveals an interesting balance between technology and growth. Technology provides increased productive capacity from available resources. The improved, technology-supported productivity precedes population growth and its accompanying demand for new resources. Trends in population growth we see in urban centers occur after the technologies needed to support these communities became available. Before efficient energies were harnessed for work and transportation of food, large urban centers were not sustainable. *Community growth was made possible by the advent of new technologies.*

Certainly, today's world has varying standards of living, from the destitute to the opulent. Volumes are filled with debate over the reasons: inequitable distribution, limited production means, oppression, lack of resources, corrupt governing structures, etc. Techonomic analysis posits that technology will arise to solve economically motivated demands when free market forces are active. Without an economically motivated cause (demand), there is no lasting technology effect (supply). But so long as a society creates an environment that enables its citizens to be rewarded for value-added efforts, then innovation flourishes and the possibility for advancement and sustainability increases.

Continuously advancing technology supports increasing population density with improving standards of living. As the population density increases, humanity becomes more interdependent, and the productivity demands on certain key technologies regulate growth. In addition to energy production, advances in agriculture

Creating Techonomic Metrics

and healthcare have historically determined the limits of possible growth, while media and religion have shaped the social mores and cultural harmony of community life. A brief techonomic look at each of these foundations of sustainable community gives further examples of how to create a techonomic metric to observe trends.

Agriculture

From the first cultivation of crops allowing nomadic cultures to settle in villages, to the recent genetic engineering of seeds, to improved yield and disease resistance, agriculture reveals a history of productivity gain. An historical techonomic metric for a given agricultural crop would combine the economic output (measured in crop quantity, bushels or pounds) with the unit labor required to produce the output (hours). One can argue that the economic value of a crop differs each year due to market production, weather conditions, etc., but in the broad techonomic view, the question is how technology affects productivity. Market choices about how much will be produced are determined by economic forces that influence how many acres are planted each year. By tracking labor productivity of a single crop throughout recent history, one can observe the impact of new technologies on productivity.

Agricultural TM — Historic

Assumptions:

1. Purpose is to observe trends in agricultural productivity as a function of the influence of technology introduction.
2. Fundamental techonomic components are agricultural production (bushels of product) per labor used (hour).

$$TM_{Agriculture\ Historic} = \text{product/unit of labor [bushels/man-hour]}$$

Agricultural labor productivity soared with the introduction of mechanization, fertilization, and genetic hybridization. Figure 4.11 shows the historical productivity gains in U.S. agriculture resulting from use of advancing technology.[22] Since the vast majority of labor content has already been eliminated, this figure demonstrates further productivity gains related to labor savings will be minimal at best. The future of agricultural productivity will be in more effective use of another limited resource: land.

Gains in yield per plant or per unit of land will determine the crops of choice in the age of genetically engineered agriculture, artificially controlled growing environments, and processing of caloric content into palatable food. A possible techonomic measure suited to comparing current methods and approaches to agricultural production would compare the key output of food (caloric content) with the limiting resources needed to produce it (land) as follows.

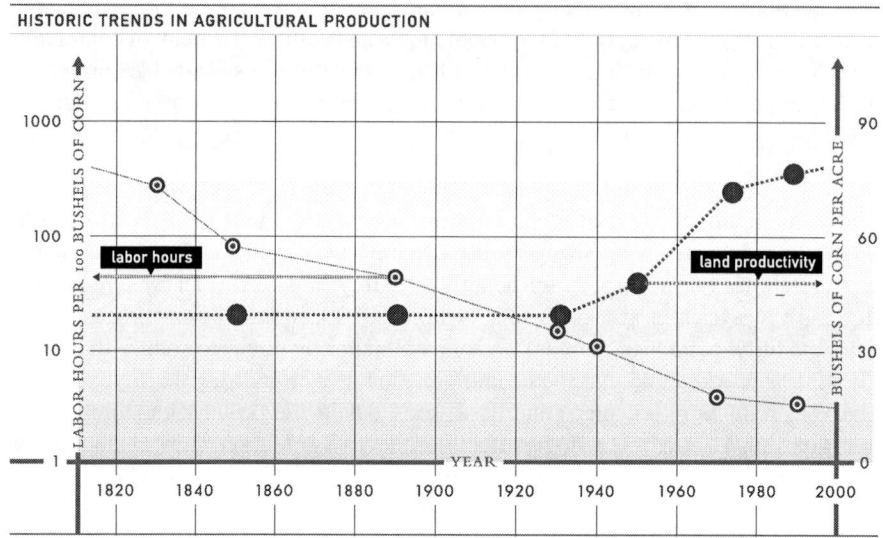

FIGURE 4.11 Historical productivity gains in U.S. agriculture resulting from advancing technology implementation.

Agricultural TM — Comparative

Assumptions:

1. Purpose is to compare current competing agricultural techniques based on their food energy value and use of land resources.
2. Fundamental techonomic components are caloric production (caloric content of edible product) per land used (acre).

$$TM_{Agriculture\ Comparative} = \text{edible energy content/unit of land [calories/acre]}$$

The comparative metric can be used to select crops for cultivation in undernourished markets, select fertilizers and seed based on their performance, or evaluate entirely new methods of cultivation like hydroponics or three-dimensional agriculture. The technology of mechanization minimized the issue of labor in agricultural production. Emerging technologies related to genetics, chemistry, germination, maturation, disease prevention, pest elimination, and harvesting will continue to advance agricultural productivity.

HEALTH AND ENVIRONMENT

One historic techonomic metric for the health of a society can be measured by the life expectancy of its members. If living conditions are improving in the environment, if food supply is plentiful and equitably distributed, and if medical practices and availability are improving, then one would anticipate a lengthening lifespan for the general populace (in the absence of war, natural catastrophe, unknown plague, etc.).

Creating Techonomic Metrics

The life expectancy (years) per person within a given group is a techonomic metric for health. This metric is useful for historical discussion, but it is of limited economic value for a contemporary analysis of the healthcare system because it lacks a fundamental economic element.

Community Health TM — Historic

Assumptions:

1. Purpose is to observe trends in quality of health over time.
2. Fundamental techonomic component is the average lifespan of the community member, the person.

$$TM_{Health\ Historic} = \text{average lifespan of community member [years/person]}$$

Figure 4.12 shows the community health quality of life metric (average lifespan) in the U.S. for the last century.[23] The significant increases in this metric in the twentieth century can primarily be attributed to improved nutrition, advanced medical practices, and reduced infant mortality.

As medical technology advances and treatments to enhance and extend life proliferate, society is increasingly faced with mounting healthcare costs. Whether rising healthcare costs are attributable to litigation, insurance, hospitals, pharmaceuticals, or practitioners is the subject of serious debate at present (see Chapter 8). Even deeper concern arises from the simple economic consideration of how much healthcare a society can afford. An interesting techonomic metric that tracks the cost

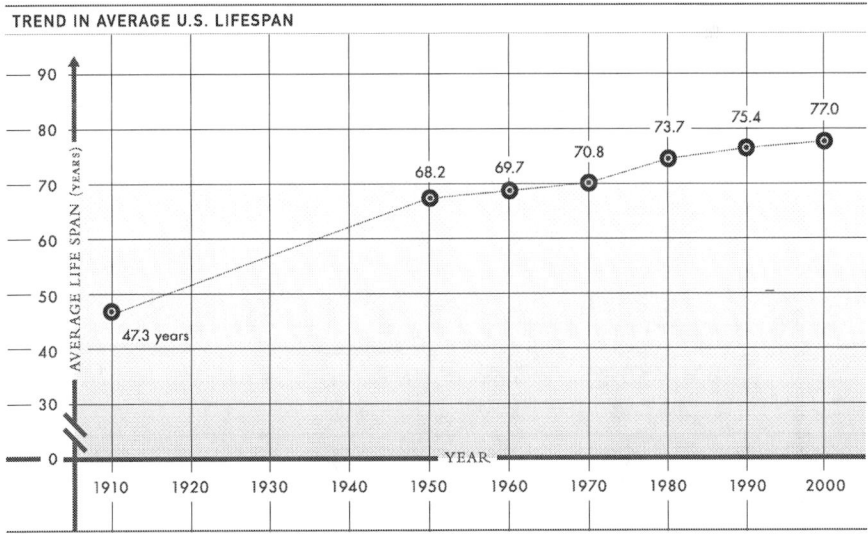

FIGURE 4.12 Community health quality of life metric, average lifespan in U.S. for the last century.

of healthcare as a result of technological advance is provided by the comparative healthcare metric.

Community Health TM — Comparative

Assumptions:

1. Purpose is to observe the economic impact of healthcare technology and provider systems.
2. Fundamental techonomic component is total dollars spent over average lifespan of an individual.

$TM_{\text{Health Comparative}} = 1/(\text{healthcare expenditure in \$ per year} \times \text{lifespan in years})$ [Life/\$]

Figure 4.13 shows the average annual per capita healthcare costs in the U.S. over the last 50 years.[24] These annual costs are increasing significantly as the lifespan also increases. The TM combines these two increasing factors to reveal a personal healthcare lifetime cost that is rapidly increasing (note that the actual TM should be inverted to be consistent with other TM combining performance, lifetime health, relative to costs, lifetime expenditures). While new technologies and processes extend the average lifespan, the cumulative cost of more frequent access to medical technology is significantly increasing the lifetime expenditure per person for healthcare. Individual costs associated with a single medical procedure are declining, while the number of procedures one experiences in a lifetime is increasing more rapidly.

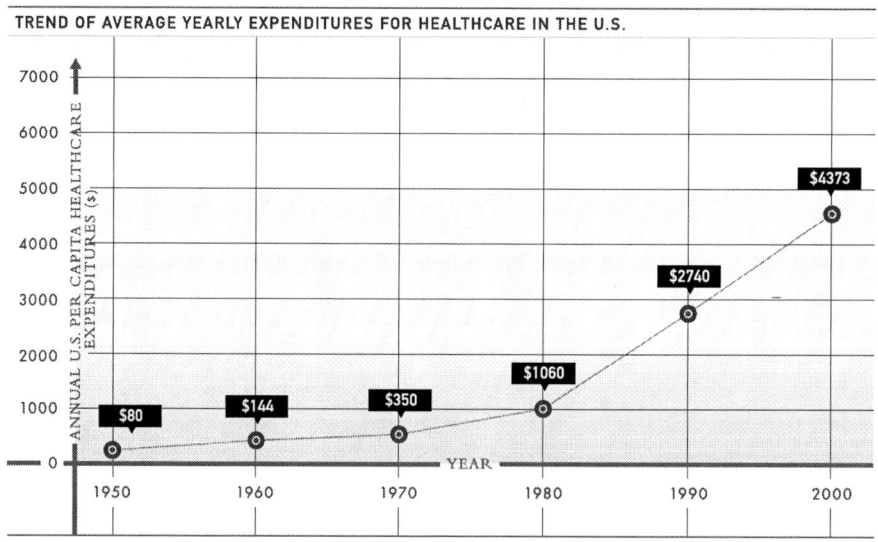

FIGURE 4.13 Average annual healthcare expenditures per capita in the U.S. over the last 50 years.

Left unchecked, this techonomic trend will overburden the societies that do not reasonably cope with the performance/cost relationship afforded by new medical procedures and pharmaceuticals. To date, our society has viewed medical technology somewhat like military technology: survival at any cost. But ultimately, the economy cannot support this unbridled expenditure requiring massive resources for incremental increases in life expectancy. The economic, moral, and ethical questions posed by this techonomic trend are among the most difficult presently facing our culture.

MEDIA/ENTERTAINMENT

The mass proliferation of computers (side 2 of the organization) and communications (side 3 of the organization) has created opportunities for shaping the habits, activities, and interests of entire cultures on a scale never before possible. *Ancient predictions of taking a message to the entire Earth have been made possible by the startling advance of space-age communication technology.* Not only can a message reach the entire Earth, it can do so, via satellite telemetry, in a fraction of a second.

Rapid proliferation of media networks has changed personal lifestyles and entire cultures. These changes are accelerating and becoming ubiquitous with new wireless communications and individualized information/entertainment access. A simple techonomic metric that shows the dramatic effect of media influence on the daily routine of the populace is constructed by tracking the average annual media usage per person.

Community Mass Media Influence TM — Historic

Assumptions:

1. Purpose is to observe trends in media influence over time.
2. Fundamental techonomic component is average annual media consumption of community member.

$TM_{\text{Media Historic}}$ = annual media consumption of community member [hours/person]

Media can be subdivided into two categories: mass (newspapers, magazines/journals, theater, films, radio, television, the World Wide Web, books, CDs, DVDs, videocassettes, audiocassettes, and other forms of publishing) and personal (personal speech, telephone, mail, e-mail, etc.). Before the technologies of the twentieth century, only books, magazines/journals, newspapers, theater, direct speech, and written mail existed as media forms. Before the printing press, the offerings were even more limited: direct speech, theater, manuscripts (hand-written texts). Figure 4.14 shows the rise in consumption of mass media usage over time.[25] Douglas Galbi postulates that the rise in media consumption in twentieth-century society is directly related to the increase in discretionary time.[26] As the workweek reduces, media consumption increases to fill the remaining time. Even if your workweek is longer, an ever-increasing number of employees have media readily accessible at the desktop via the Internet, television, or radio.

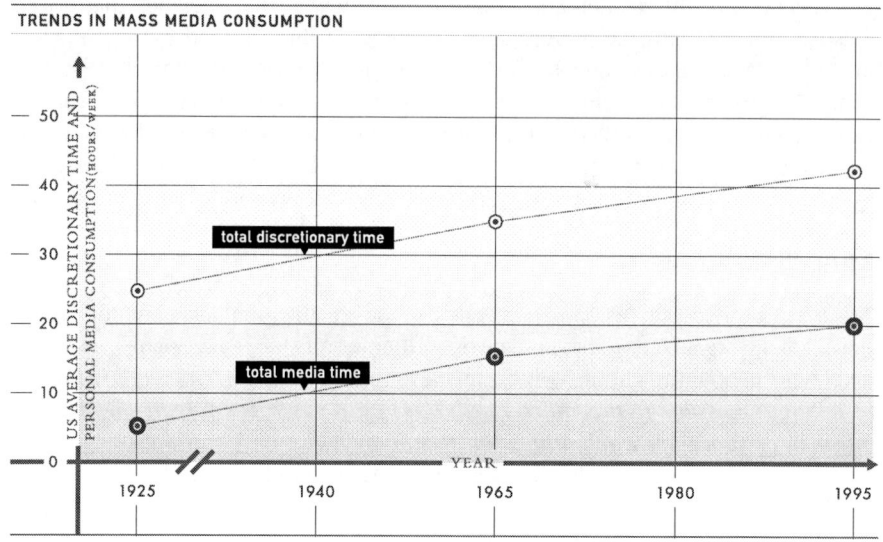

FIGURE 4.14 Rise in consumption of mass media usage over time.

One can globally track total advertising expenditures per media type or track average usage hours per media type to compare the current trends in media preference and hence the future value. Data for usage patterns from surveys is more easily obtained than total industry advertising expenditures by media segment. Comparative techonomic measurement of mass media therefore focuses on societal usage patterns rather than attempting to track segmented media costs.

Community Mass Media Influence TM — Comparative

Assumptions:

1. Purpose is to observe the trends in current mass media influence for different media types.
2. Fundamental techonomic component is average annual media consumption of community member for each competing media type.

$TM_{\text{Media Comparative}}$ = annual media consumption of community member [hours/person]

Note that the historic and comparative techonomic measures are the same equation. But they are applied in completely different ways. Historically, observing all mass media usage, one observes the direct influence of technology on mass media influence and societal impact. Comparatively, observing mass media preferences of the population over a short period of time, the shifting preferences can be identified and used to advantage — same equation, different inputs, different trends observed. Keep in mind that the goal of techonomics is to provide a predictive metric for the future direction of a competitive market, enabling wise deployment of resources.

Predicting trends is much different, and arguably more valuable, than predicting absolute quantities.

RELIGION OR "SPIRITUAL" CONDITION

How does one measure the vitality of an organization or society, particularly without infusing his or her personal definition of spirituality into the metric? Security vs. freedom, order vs. chaos, absolutes vs. relatives, what is good for the whole vs. what is good for the individual, rule of law vs. evolution of law, material wealth vs. spiritual wealth. How do you measure something you cannot touch, see, or quantify, but something your gut tells you is there? Whether or not you believe an individual has a "spirit," you can certainly observe the spirit of a corporate culture or the spirit of national patriotism.

The vigor of an organization, in some sense related to its "spiritual" condition, can be measured by its rate of growth or decline. Two fundamental measures of an organization's vitality are headcount and revenues. From Main Street to Wall Street, growth in profits or revenues is the hallmark of the condition of a company. These numbers are easily manipulated in today's complex accounting environment (Enron, WorldCom, others), so they must be carefully scrutinized. On a national level, gross national product or immigration patterns give an indication of societal economic health, but still no insight into the spiritual condition of the culture.

This is a question for all of us to keep asking ourselves, throughout life. How do we measure the spiritual health of an individual or an organization, a company or a nation? It is a particularly compelling question for the leader, who must be concerned with establishing and inculcating a viable corporate culture.

REFLECTIONS ON INTERDEPENDENCE

The technology timeline indicates humanity is becoming more specialized and increasingly interdependent. We are all more dependent on each other today than ever before in history. The reason is advancing technology deriving efficiency from specialization, not generalization. The main impetus in this direction came from harnessing energy, entering the industrial age, and specialization of our work efforts. Now we rely on the megafarm for our food, and most of us do not even know how to milk a cow or slaughter a hog or bake bread.

This state of affairs has its up side. When more people are freed to do intellectual work, and can collaborate more easily, technologies tend to grow, and cooperative humanity is blessed with better medical care, food production, living conditions, and standard of living. But interdependence has its dangers as well. Ponder the consequences of losing our utilities (electric power, fuel, water, or telephone) for an hour, a day, a weekend, a week, a month. How long does it take for chaos to develop? An hour without power is just an inconvenience; read a book by candlelight or even talk to your family. A day starts to get worrisome. The food in the refrigerator is starting to thaw, and the hot water tank is cooling as the house reaches ambient temperature. A weekend is now troublesome. It is time to leave and find somewhere with air conditioning, lights, operational gas pumps, food preparation, communications (TV, Internet, radio). If the electrical outage continues in a widespread manner for a week, we are now desperate for food, physical comfort, and assurance that things are being brought back to normal. I do not know if most unprepared urbanites could make it for as long as a month. I hope I never have to find out, although the tragedy of Hurricane Katrina offered a glimpse into these possibilities. Our modern urban infrastructure has become so reliable as to be taken as an entitlement. It was established and is perpetuated by the cooperative expertise and efforts of many.

Unlike the past, we are users of technologies and networks that individual users cannot maintain. Today, technology products and systems are the culmination of the efforts of many minds and resources, but they are only superficially understood by their users. If the computer crashes, all most know to do is to reboot. If the car will not start and the problem is more complicated than fuel or a battery, it is time to call the tow truck. If the TV is inoperable, it is easier to buy another than try to diagnose and repair it. Combine that thought with the design strategy of planned obsolescence (a design approach that considers product failure or future feature advancement as an opportunity to create repeat customers). You end up with a populace running a constant race to keep more and more things operational that are less and less reliable. At times this situation becomes most frustrating!

Our society goes to great lengths to make things inordinately easy. Velcro replaces the skill of tying shoes. Automatic sensors turn on our faucets and flush our toilets in public places. Food is preprocessed so that it can go directly from refrigerator to the table with only an unskilled pass through the microwave. Constant media access replaces the quiet imagination. The calculator and the cash register perform all of our mathematics without our need for the skill. The need to learn spelling is superseded by the ever-present spell-checker, and fortunately, the grammar checker catches most egregious spelling errors.

Still — most of us of seem to think it is worth it. **We are the beneficiaries of many and vast collaborations.** Technology is raising the standard of living comfort level so high that the natural struggle that strengthens character, physique, and mind is being diminished in our culture. Are we mentally more capable as our interdependence avails more free time? Are there rational limits to dependency? The metaphor of the animal within a zoo where all needs are provided may be pleasant for a generation, but it may compromise foundational skills to the point that survival without modern conveniences is threatened.

On a positive note, because of the advance of technology, even the poorest in our society experience creature comforts that surpass all who lived on Earth before the nineteenth century. Our society as a whole is more aware and able to prepare, adapt, and respond to unexpected challenges (storms, disease, etc.) than ever before in history. The available energy augmenting human physical capability, combined with instantaneous global communications and expanding mental leverage, creates opportunities for community growth that could never have been imagined by those preceding us. Techonomics provides a key to unlocking those opportunities by discerning judicious utilization of precious financial and human resources.

SUMMARY

TECHONOMIC METRIC

A metric is a measurement that combines a technological capability with a cost component to create a value that can be monitored with available data. Cost can be a function of price or a function of labor content for activities that predate current monetary valuations. Techonomic metrics are valuable in the rational analysis of trends affecting a market.

LEADING INDICATOR

Military developments lead the mass-market deployment of technology, because military research is performed without the intense financial constraints of the commercial market. Military technology development is supported by the cultural will to survive at all costs.

ENERGY

Before steam power, human energy consumption was limited to wood/peat/coal for cooking and heating, and to animate power for construction and cultivation. The steam engine and subsequent technologies to harness fuels have resulted in waves of increasing energy consumption over the last four centuries. We have tapped fuels of increasing energy density or portability. Electric power allowed for central generation and remote distribution, connecting large segments of humanity with large amounts of power.

COMPUTATION

Electricity transformed computation from mental and mechanical to electronic augmentation. Technology further accelerated computation in the form of the microprocessor, following Moore's Law of increasing computational capability. Computational magnification has affected organizations on every level worldwide, and these effects are only in the early stages of implementation.

COMMUNICATION

Digital communication based on worldwide networking of computers is transforming organizations and their modes of operation. Globalization of operations is encouraged and sustained by the transparency and low cost of communications as this network continues to expand.

COMMUNITY

At the same time technologies are allowing greater urbanization, communities are also migrating from real to virtual. As communities grow, greater specialization and interdependence mark their members.

Key Terms

historic techonomic metric
comparative techonomic metric
interdependency
performance/cost ratio
techonomic sweetspot

QUESTIONS

1. Techonomic metrics combine a technological measure and an economic one. The digital camera metric is simply PIXELS/COST. Locate the technical specifications for 2 to 4 digital cameras on the market (Internet or visit a local electronics store), and calculate the digital camera techonomic metric for each. Certainly, other considerations affect the buying choice (size, ease of use, compatibility, supporting software, etc.), but this key techonomic metric will provide a first approximation to the value proposition of the comparison products.
2. Referring to Question 1 and Figure 4.2, what trend do you anticipate the digital camera techonomic metric will follow: (A) continue to increase rapidly, (B) continue to increase more slowly, (C) level out, (D) other? Explain your choice.
3. Figure 4.3 reveals a marked decline in film camera sales in recent years. This is evidence of the journey from "atoms to bits" that Negroponte has written about: atoms (film) are being replaced by bits (digital memory). List one or more products/markets where you observe this migration from atoms to bits.
4. Military developments are the leading indicators of technology realization. Create a list of technology-based products you use today that grew from military development. Now, anticipate at least one technology now in military use that you predict will become common in the mass market in years ahead.
5. The performance/cost sweetspot is an important concept to the discernment of new market opportunities. Many ideas save some money or add a new performance benefit. The performance/cost sweetspot seeks those technologies that offer the potential to both reduce costs and add unique performance benefits. One interesting technology that is emerging is Voice Over Internet Protocol (VOIP). VOIP offers very low-cost telephone communication, but must overcome a reputation for limited quality and user friendliness. Place VOIP on the cost/quality chart, and discuss keys to mass-market penetration of this emerging telephony approach.
6. New energy conversion technologies have created increasing demands for fuel, affecting both the amount and type of fuels used. Wood for cooking/heating fires served humanity for thousands of years. Next, coal fueled the steam engine (external combustion) and, with it, the rise of the industrial age. A century later, petroleum fueled the mobility provided by the internal combustion engine. (A) Do you accept or reject the premise that

standard of living is greatly affected by per capita energy consumption? (B) Given technical, environmental, economic, and social constraints, what scenario would you predict for U.S. energy production in a 10-, 20-, or 40-year timeframe?
7. Computation has rapidly accelerated ever since humanity captured and exploited electricity (the last century). Before electricity, only human-initiated mechanical means (abacus, slide rule, etc.) were available to leverage computational capability. In the last 40 years, the microprocessor has further accelerated these trends. If Moore's Law continues (see Figure 4.6), the desktop computer will outpace human memory, computation, and inference by the year 2020. Discuss the implications of this to your career, profession, or organization.
8. Communication capacity and proliferation has expanded as computational advances have facilitated the transition from analog to digital media. This expansion has been most pronounced in the last 20 years, with the advent of the Internet and its ability to deliver data packets of any kind of digital media nearly instantaneously worldwide. Communications around the world are now as easy and inexpensive as communications across town. Discuss the implications of this to your profession or organization.
9. Communities are now able to form "virtually." Like-minded individuals can pursue activities of common interest without physically meeting. Create a list of activities that you used to do physically but now do virtually, OR describe a virtual community that you participate in (eBay, adventure games, political action, etc.).

REFERENCES

1. Delis, D., Total camera sales, Charts for speakers, Photo Marketing Association, March 4, 2005, http://www.pmai.org.
2. Smil, V., *Energy in World History,* Westview Press, Cambridge, 1992, p. 239.
3. Smil, V., *Energy in World History,* Westview Press, Cambridge, 1992, p. 236.
4. National Energy Information Center, International Energy Price Administration, Energy International Administration, October 12, 2005, http://www.eia.doe.gov/emeu/international/prices.html.
5. HearthNet, Fuel Cost Comparison Calculator, HearthNet, 2005, http://www.hearth.com/fuelcalc/findoil.html.
6. World Nuclear Association, Energy for the world — Why uranium? World Nuclear Association, 2001, http://www.world-nuclear.org/education/whyu.htm.
7. Energy Information Agency, Nuclear power in France, Green Nature, April 2003, http://www.greennature.com/article744.html.
8. White, S., A brief history of computing — complete timeline, Oxford University Computer Society, 2005, http://www.ox.compsoc.net/~swhite/history/timeline.html.
9. Wikipedia, ENIAC, Wikipedia, The Fee Encyclopedia, 2005, http://en.wikipedia.org/wiki/eniac.
10. Kurzweil, R., *The Age of Spiritual Machines,* Penguin Group, New York, 2000.
11. Smolan, R., *One Digital Day: How the Microchip is Changing Our World,* Crown Business, New York, 1998.

12. Darlington, R., Fascinating Facts and Figures about all aspects of the Information Society, Roger Darlington's Homepage, 2005, http://www.rogerdarlington.co.uk/FFF.html#Books.
13. Drye, M., Vellum replaces papyrus, ancient book-making of the early Roman world, University of North Carolina at Chapel Hill, 1997, http://www.unc.edu/courses/rometech/public/content/arts_and_crafts/Meredith_Drye/Ancient_Bookmaking.htm#VELLUM%20REPLACES%20PAPYRUS.
14. Institute of Paper Science and Technology, The spread of papermaking in Europe, Robert Williams Paper Museum, Institute of Paper Science and Technology at Georgia Tech, 2005, http://www.ipst.gatech.edu/amp/collection/museum_pm_euro.htm.
15. Regents of the University of California, How much information? Regents of the University of California, 2003, http://www.sims.berkeley.edu:8000/research/projects/how-much-info-2003/execsum.htm.
16. Tanton, J.H., End of the migration epoch, *The Social Contract Press,* IV, 3, 1995, http://desip.igc.org/populationmaps.html.
17. U.S. Census Bureau, Historical Estimates of World Populations, United States Census 2000, U.S. Census Bureau, 2000, http://www.census.gov/ipc/www/worldhis.html.
18. Chandler, T., *Four Thousand Years of Urban Growth: An Historical Census,* St. David's University Press, 1987.
19. Rosenberg, M.T., Largest cities through history, About Inc, 2005, http://geography.about.com/library/weekly/aa011201a.htm.
20. Mongabay.com, Urban agglomerations (city urban areas) with more than 1 million people in 2003, population stats, Mongabay.com, 2004, http://www.mongabay.com/igapo/2005_world_city_populations/2005_urban_01.html.
21. Wikipedia, List of countries by population density, Wikipedia, The Fee Encyclopedia, 2005, http://en.wikipedia.org/wiki/List_of_countries_by_population_density.
22. Bellis, M., A history of american agriculture, Inventors, About Inc, 2005, http://inventors.about.com/library/inventors/blfarm1.htm.
23. Wikipedia, Life expectancy over human history, Life Expectancy, Wikipedia, The Free Encyclopedia, 2005, http://en.wikipedia.org/wiki/Life_span#Life_expectancy_over_human_history.
24. *McBride, T.D.,* Is the sky falling? Rising health care spending and planning for the future, Rural Health Panel, Rural Policy Research Institute, 2005, http://www.rupri.org/ruralHealth/presentations/McBride_NRHA_Jul20_v3.pdf.
25. Baker, F.W., Media use statistics: Resources on media habits of children, youth and adults, Media Literacy Clearinghouse, 2005, http://medialit.med.sc.edu/mediause.htm.
26. Galbi, D.A., Some economics of personal activity and implications for the digital economy, *First Monday,* 6, 7, 2001, http://www.firstmonday.org/issues/issue6_7/galbi/index.html.

Section 3

Techonomics at the Turn of the Twenty-First Century

5 The First Three Laws of Twenty-First-Century Techonomics

Make things as simple as possible, but not more so.

— Albert Einstein

INTRODUCTION

The first three laws of twenty-first-century techonomics develop a philosophy of progressing organizational efficiency that is broadly applicable and personally observable. These three "laws" form the foundation of twenty-first-century techonomics, because these emerging technology trends are the most powerful economic drivers reshaping today's organizations. The first three laws of techonomics are:

- **First Law — Moore's Law (Law of Ubiquitous Computing):** Named for Gordon Moore, a founder of Intel, this widely known and validated observation describes the diminishing cost of electronics due to technological advance. Loosely stated, Moore's Law predicts that the cost of electronic computation is cut in half every 18 months.[1] Ubiquitous computing results as the cost for computing capability diminishes to the point that computers are included in virtually every manufactured device.
- **Second Law — Metcalfe's Law (Law of the Ubiquitous Global Network):** As the inventor of Ethernet, one of the first practical computer networking systems, Bob Metcalfe observed the exponential growth of connections between computers as the number of networked computers increased.[2] This growth has moved from theory to practice with the rise of the World Wide Web. Simply stated, as the number of computers on a network increases, the number of interconnections increases exponentially. The law of the ubiquitous network predicts that the global network will rapidly expand, and its access will become universal.
- **Third Law — Coase-Downes-Mui Law (Law of Diminishing Organization Size):** As a twentieth-century economist and philosopher, Ronald Coase developed the theory of transaction cost analysis, the "make-or-buy" decision.[3] Propelled by access to near perfect information provided by the ubiquitous network, Coase's transaction analysis tends to the "buy"

decision. Downes and Mui predict a significant rise in outsourcing as firms seek the most effective production means through external suppliers.[4] A firm's employee count continually diminishes as production shifts to the most efficient suppliers, hence the Law of Diminishing Organization Size. This law can also be termed the Law of Increasing Productivity.

Techonomics incorporates these laws into a method for analyzing significant trends that affect organizations throughout business and society.

In this chapter, these key twenty-first-century techonomic laws are described, examples of their validity are given, and their direct effects on organizations are discussed. However, techonomics does not end with these three "laws," it only begins there. Your challenge and opportunity is to learn how to observe and extend these analysis methods to your own organizations and endeavors.

MOORE'S LAW: UBIQUITOUS COMPUTING

FIRST LAW OF TECHONOMICS ADAPTED FROM MOORE'S LAW (LAW OF UBIQUITOUS COMPUTING)

The cost for equivalent computing performance halves every 18 months,

OR conversely,

Computational performance at a constant cost doubles every 18 months.

Gordon Moore, cofounder of Intel, observed in 1965 that the number of transistors per square inch on an integrated circuit had doubled almost every year. Moore predicted that this trend would continue for some time into the future. He predicted that the density of electronic components in an integrated circuit will double every 18 months for an indefinite period of time. In the four decades since Moore's prediction, the timeframe for this density doubling has been measured consistently in the range of 18 to 24 months, and this trend appears not only to be continuing, but to be tending closer to the 18-month time frame. Since the gain in density is geometric (it doubles) in a linear time frame, the compounded results over time are remarkable. The First Law of Techonomics is called the Law of Ubiquitous Computing because *the exponential trends in performance improvement and cost reduction for electronics at some point makes it economically feasible to put computational capabilities into virtually every manufactured item, from tennis shoes to toothbrushes.* Computing capabilities are embedded in almost every conceivable product.

Figure 5.1 shows evidence of Moore's Law over three decades: the number of transistors in the latest microprocessors from Intel over the last 30 years. Moore's Law has been a strong predictor of electronic density throughout this period. Moore's Law is modified from a technology law to a techonomic law when additional technical and economic observations are used. ***Both speed and cost of an integrated***

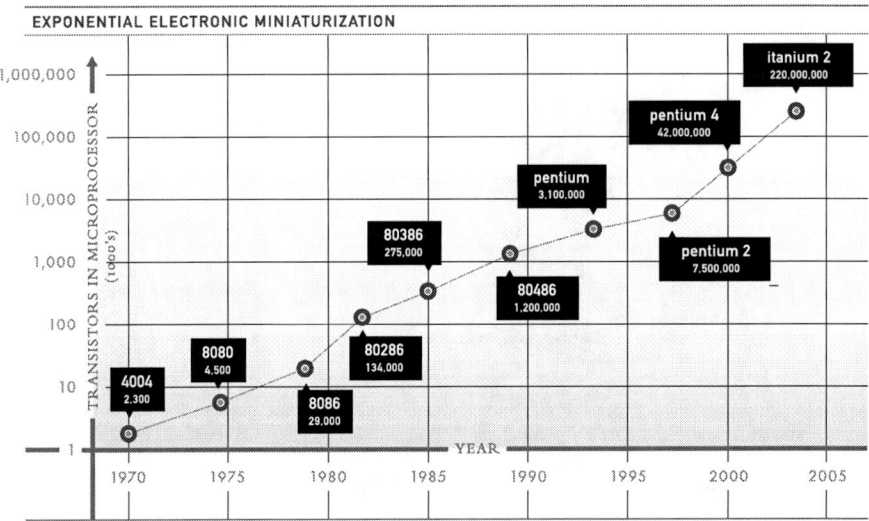

FIGURE 5.1 Evidence of Moore's Law over 30 years.

circuit are tied to its packing density. The primary economic cost to an integrated circuit is the wafer on which the circuit is printed. Cost of these wafers is mostly a function of size, so if the wafers stay at the same size, wafer cost is fixed. The smaller the transistor size, the more transistor elements can fit on the same wafer size. The result: more product (computational speed) for the same price as transistors get smaller. That is why Moore's Law can also be stated as: Computer performance doubles at the same cost or its cost halves for the same performance every 18 months.

There are several easily understood implications of the First Law of Techonomics (Moore's Law Modified). Since the performance of smart electronic devices relentlessly doubles every 18 months, what was impossible yesterday (due to either size, cost, speed, power requirements, or intelligence) becomes possible today and commonplace tomorrow. For example, in 1969, the computer that augmented control of the Lunar Excursion Module (LEM) that placed men on the moon and returned them safely to the mother ship had less computational power than the PDA (Personal Digital Assistant) widely available and affordable today. Figure 5.2 shows the implications of exponential advance in computation with accompanying diminished cost — 30 years from moon to pocket. Not only was the LEM a little large to carry in your pocket, but it was pricy for the typical consumer.

Over the last 40 years, the implications of Moore's Law have permeated almost every consumer product imaginable. According to a page on the Intel website celebrating the 40th anniversary of Moore's Law, because of the relentless technological and economic progress realized by electronic miniaturization, it is now cheaper to make a transistor than it is to print a single character in a newspaper. It is little wonder newspapers are becoming economically challenged by electronic means of information distribution. From smart traffic lights, home lighting controls, automobile windshield wipers, automated telephone systems, and digital watches, myriads of products now think for themselves. This massive distribution of imbedded

THIRTY YEARS FROM MOON TO POCKET

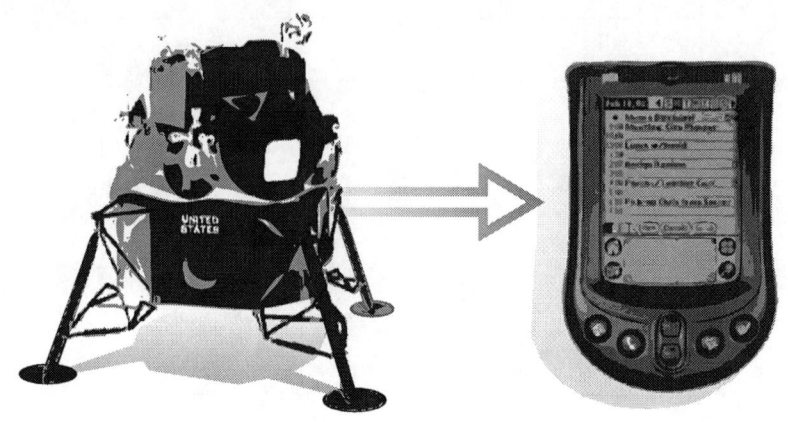

FIGURE 5.2 Implications of exponential advance in computation with accompanying diminishing cost.

computation has been made possible, not just by technological advance, but by the economic viability of the technology — the *techonomics*. Many new automobiles contain upwards of 50 embedded microprocessors, and the typical American home contains hundreds of them in products and appliances.

Recall Kurzweil's prediction in *The Age of Spiritual Machines:* by 2040, a $1,000 PC will hold the knowledge of the entire living human race, and its analytical powers will be enormous.[5] This is the implication of the progression of Moore's Law, and it is staggering.

First Law Implications

- Shortened electronic product life cycles.
- Ever increasing number of "smart" products.
- Machine intelligence (automation) less expensive than human labor for significantly increasing number of occupations, particularly bureaucratic labor.
- Intellectual expertise can be captured and transferred at progressively diminishing costs.

METCALFE'S LAW: UBIQUITOUS GLOBAL NETWORK

SECOND LAW OF TECHONOMICS ADAPTED FROM METCALFE'S LAW (LAW OF THE UBIQUITOUS GLOBAL NETWORK)

> The cost of reaching anyone or finding anything on the network diminishes exponentially with the number of users.

Robert Metcalfe was the inventor of the Ethernet protocol, an early and successful method for joining computers into a network. He was also a founder of 3Com. He made the observation, now known as Metcalfe's Law, that *the connections of a network increase in proportion to the square of the number of nodes.*

Figure 5.3 provides a visual representation and mathematical equation demonstrating Metcalfe's Law. Since there is great value in being connected, Metcalfe's Law has been commonly modified as: *The value of a network increases in proportion to the square of the number of users.*

Certainly, AT&T (American Telephone and Telegraph) prior to deregulation would have concurred with Metcalf's Law. For nearly 100 years, each new customer installation made their telephone line utility both more valuable and a greater barrier to competitive entry. Deregulation of the industry and the emergence of the wireless network removed the protection of their proprietary network.

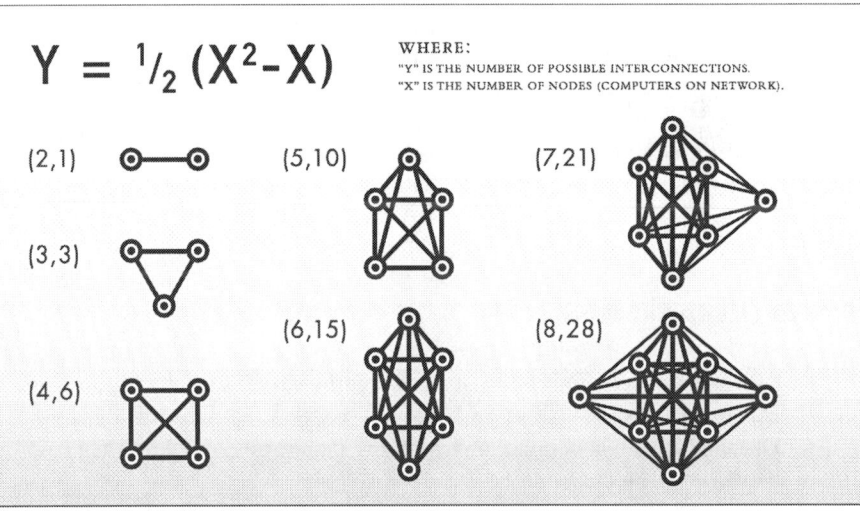

FIGURE 5.3 Graphic representation and mathematical equation demonstrating Metcalfe's Law.

Over the past decade, choices of telephone carriers and methods have proliferated, and the value of the AT&T network has been significantly diminished. When we use the Internet to send email or search for information, we are generally not even aware of whose network we are accessing, though we may know the portal through which we entered.

Given these observations about the diminishing value of the network as open systems compete, what is the techonomic perspective on Metcalfe's Law? The essential question is: what is the economic impact of the obvious technological creation of a global network? Near perfect access to information. *The cost of reaching anyone or finding anything on the network diminishes exponentially with the number of users.* One can now reach or find most anything, or anyone, from almost anywhere, almost instantly.

The combination of Metcalfe's Law and Moore's Law has accelerated the spread of the global network to warp speed. Kurzweil's "Law of Accelerating Returns" (incremental additions result in increasing returns) is the twenty-first-century inverse of the traditional "Law of Diminishing Returns" (incremental additions result in diminishing returns). Unlike the valuable telephone network that allowed for linear conversations among a few captive parties, digital information is accessible via the network in real time as well as massively storable and searchable via intelligent methods. With over eight billion pages cataloged by Google, you do not even have to know where to look. No more linear searches through microfilm libraries, no more mass mailings to incorrect addresses, no more dead-end telephone calls following leads.

In less than a decade, the world has virtually shrunk to a desktop; virtual access to the world's information is at your fingertips. And the cost of this access — measured in the savings of time, postage, false dead ends, telecommunications cost, access cost, etc. — has plummeted as technology has moved transactions from the tangible and time-bound world (travel, postal mail, linear telephone calls) to the virtual world (Internet, search engines, ubiquitous computing, chat rooms, nonlinear information retrieval, virtual conferencing). Truly, if knowledge is power, we are the most powerful people ever to inhabit the Earth.

SECOND LAW IMPLICATIONS

- Access to people and information is significantly easier, faster, and less expensive than ever before in history — near perfect information.
- Distance has no economic impact on the monitoring and performance of endeavors — globalization.
- Exponentially more information is being generated and transmitted, much of it without human intervention — nonlinear access.
- The network is the foundation of twenty-first-century commerce — just-in-time processes.

COASE-DOWNES-MUI LAW: DIMINISHING ORGANIZATION SIZE

THIRD LAW OF TECHONOMICS: COASE-DOWNES-MUI LAW (LAW OF DIMINISHING ORGANIZATION SIZE)

As transaction costs diminish, smaller organizations evolve.

All things are intertwingled.

— Leo Brodie

In their analysis of how the Internet has impacted transaction costs, Larry Downes and Chunka Mui extended the transaction cost analysis of British economist Ronald Coase. Their conclusion: *As transaction costs diminish, smaller organizations evolve.* This observation will be identified as the Third Law of Techonomics: *The Law of Diminishing Organization Size.* **As technology has supplied more and more "perfect information," reducing transaction costs and risks, it has become easier to outsource more of an organization's work to efficient and reliable sources, reducing the need for internal company employment.**

Let us go back to the phrase "transaction cost analysis." Ronald Coase first described this kind of analysis in a 1937 article entitled "The Nature of the Firm."[6] In layman's terms, transaction cost analysis is simply the *make-or-buy* decision. Should I make a needed product or buy it from some external source? As individual consumers of goods, we perform these make-or-buy decisions many times daily, usually opting to buy more often than make, as our lives become more mutually dependant. Corporations, likewise, continually analyze make-or-buy transactions in order to compete economically in the free market.

Coase found that many elements go into transaction analysis once a need is determined. A list of some of these transaction elements defined previously in Chapter 2 follows:

- Availability
- Quality
- Transport
- Punctuality
- Inventory
- Switching
- Risk
- Availability
- Quality
- Reliability
- Price
- Trust

Every day, people all over the world exchange their time for money and their money for needed commodities. Each generation makes these decisions differently, based on their skills, the availability of needed products, and the demands on their time. The last 200 years in the U.S. have witnessed a huge shift in personal outsourcing as we have moved from an agrarian society to an industrial one.

Coase studied several U.S. companies in the early 1930s. In particular at Ford Motor Company's River Rouge Plant, he found a monolithic (self-contained) factory that turned iron ore into automobiles. Since Ford had pioneered mass manufacturing, obtaining parts in the quantities needed to meet production was hard to do. He was forced to make, rather than buy, many components.

The automobile industry in America went mostly with the make option for many years, even though buy options were arising. If you follow the industry's progress through the years, you see a major shift in structural organization occurring in the 1970s. This had a lot do with the Japanese, who used outsourcing effectively to improve quality and reduce costs simultaneously. The entire industry has now followed suit in order to compete in the international marketplace.

If you look at the contributing factors to the make-or-buy decision, you will see that many of them are rooted in information. The better the information, the more accessible the information, the more timely the information, the easier it is to make a justifiable value judgment to make or buy. More information leads to awareness of more options.

Consider again the automobile maker. In the 1930s, there may have been only one or two bearing makers to provide quality wheel bearings in the quantities needed for mass-producing automobiles; same for transmissions, windshield glass, headlights, etc. An automobile manufacturer would find a key supplier (if any), determine price possibilities, and then decide whether to do the job inside or farm it out. The decision was often to do the job within the company.

Now fast forward to the 1990s. The Internet is introduced. Automation and mass manufacturing have become widely practiced skills. Virtually every component on an automobile can be made in the quantities needed by a dozen or more suppliers. Now the Internet provides a means to find these multiple sources, monitor their work product, simultaneously pit them against each other to supply the same product, switch seamlessly from one supplier to another, and transact all the financial arrangements to the benefit of the buyer. No surprise that the make-or-buy pendulum has swung far to the buy side. This is also true on our personal endeavors as we become more consumer-driven and interdependent.

If organizations are getting smaller due to outsourcing many jobs previously done "in house," does this mean that jobs are dwindling away? Not necessarily. The Third Law of Techonomics says there will be smaller companies, not fewer jobs. It means there will be more opportunities for companies with a single-minded, specialized, "value-add" purpose. These companies (if they are to evolve and thrive) will provide their customers with exceptional goods and services, focusing on the things they do best, the things that cannot easily be outsourced.

Several modern business practices have arisen due to the combination of "perfect information" and transaction cost reduction. These include:

- *Outsourcing.* This is the common practice of buying products from an external source rather than making them internally. As technology advances to provide information on suppliers and the ability to remotely monitor their production rates and quantities, outsourcing increases.
- *JIT (just in time) manufacturing/delivery.* JIT manufacturing networks result from the coordination of production needs between different suppliers to minimize work in progress and significantly reduce inventory costs. Computer technology (Moore's Law) combined with expanding networks (Metcalf's Law) makes JIT possible. And with the continuing cost reductions in the related technologies, JIT now affords a strong competitive advantage to those using it.
- *Mass customization.* Mass customization takes JIT to an extreme by maintaining no inventory until the customer has placed a definitive order. Using the Internet to take orders and JIT to then fulfill them, the production pipeline holds no wasted inventory, and the producer can always use the most advanced (or inexpensive) components to supply the customer. Computer technology, combined with expanding networks and the ease of using the Internet, makes mass customization a reality today.
- *Globalization.* Outsourcing was a local approach in the 1960s, a national approach in the 1970s and 1980s, and has progressed to an international approach in the 1990s and 2000s. The diminishing cost of worldwide telecommunications brought on by the tremendous expansion of these systems in the 1990s (the Internet telecom boom) has eliminated the cost barrier for international communication. The oceans are crisscrossed with fiber optic cables, and the skies (from 100 to 50,000 miles out!) are covered with satellites. As a result, your PC service call can be less expensively handled in India than in Indiana when the total cost of labor and communications is considered (not to mention the increasing labor cost of benefits). Globalization started with manufacturing and has now progressed to service/telemarketing and software development due to the increasing interconnectivity at lower costs (First and Second Laws). Technology now allows the economics of worldwide personal commerce to flourish. As a result, anticipate labor rates for transportable occupations to level out worldwide in the generation ahead as competitive industries empower the third world populace to embrace technology and participate in the world economy. As Thomas Friedman correctly asserts, the world is (becoming) flat.[7]
- *Lean organization.* The lean organization results from minimizing a company's internal operations and layers of management, focusing on the core value proposition of the business, and outsourcing all other activities. Successful lean organizations understand the true nature of their "value add," the special quality of a good or service that makes them desired by the consumer.

Today, we send an e-mail to Singapore, look at a Web site hosted in the U.K., buy a product made in China, and get technical help from a service representative in India — all in the same day or even a few minutes. ***Instantaneous, inexpensive, and ubiquitous worldwide communications are transforming the world economic order within a single generation.*** The implications not only affect individual businesses, but marketplaces, national economies, and entire global cultures.

Another way of looking at the Third Law of Techonomics for organizations is in terms of productivity: output per employee. In simplest terms for a business, this metric becomes revenues per employee. Whether the revenue generator is a product or service, this metric is one that can be used to quickly determine current productivity and annual trends in operating efficiency. As the value of this metric increases, either the headcount is diminishing or revenues are increasing, or both. Improving productivity is the trend predicted by the Third Law.

As this simple metric is applied to a spectrum of organizations, one should observe a clustering of the values depending on the type of business, its capital requirements, and its product offerings. For instance, a cleaning service with non-skilled workers, little capital requirement, and a limited marketing overhead might thrive with an annual revenue metric of $50,000/employee. On the contrary, a large manufacturer with a significant cost of goods, high capital requirements, mass marketing costs, and distribution costs might be failing with a revenue/employee figure of $100,000. There are different ranges of this metric for different industries, but the annual trends in revenue per employee should always be increasing if the organization is thriving and adopting effective practices to increase productivity.

In view of the many opportunities for productivity gain, The Third Law of Techonomics predicts that, if your organization is not continually getting more efficient and productive, a competitor in your industry is likely to be overtaking it.

THIRD LAW IMPLICATIONS

- A relentless push for efficiency gains — increased outsourcing and lean organizations.
- Organizational success tied closely to the quality of its network of partners.
- Organization size will diminish for the same output as productivity increases.
- Growth in individual and organizational specialization and interdependence.

THE FRANCHISE EFFECT: GROWTH THROUGH REPLICATION

Here is a riddle: what is both lean and huge? Answer: a lean, widely franchised organization. Such huge organizations do not defy the Third Law of Techonomics; they are simply an effective variation of it. The most successful franchises take extensive advantage of technology, allowing them to focus on their special offering of goods and services. Each operation reduces transaction costs by collective purchasing, and they judiciously outsource many activities to a central service core. So they have learned how to become lean and grow simultaneously through distributed geographic replication of efficient endeavors.

The geographically distributed franchise was pioneered in the hospitality industry (restaurants and hotels). Geographically distributed franchises epitomize the "think globally, act locally" philosophy. They find a need, figure out the business model, fill the need, and then rapidly replicate their success. Each unit in the organization — a storefront or hotel or restaurant — gets leaner and more focused on its product/service, while the total organization grows by replication (more units are created). Franchise organizations can grow very rapidly when they take advantage of the technology networks now available to connect and manage them.

In addition to restaurants and hotels, many other single-purpose, distributed operations can be viewed as a form of franchise, whether owned in whole by a central company or in part by owner-operators. *The foundation of franchise success is to understand the key to your value and be the best at providing it.* A secret formula (KFC, Coke), an efficient infrastructure (Federal Express, United Parcel Service), a well-financed branch structure (Citicorp, Merrill Lynch), a high-quality service (Kinko's, ServiceMaster), a low-price merchandiser (Wal-Mart), a focused market leader (Walgreens, Best Buy, Staples), or any number of business catalysts now form healthy, expanding units within a larger entity.

The techonomic advantages of the Franchise Effect include:

1. Combined purchasing power.
2. Centralized development of key systems and support technology.
3. Tested financial models and business processes.
4. Distributed operations responsibility and shared financial rewards.
5. Increasing market barriers to entry for competitors.

These advantages contribute to the rapid rise and proliferation of the franchise model in a time when "right-sizing" is dominating the economic landscape of large, monolithic companies (General Motors, Xerox).

Franchises outsource many of their functions to suppliers or to a central office. Take, for example, a restaurant franchise. Restaurant architectural design, signage, and decorations are controlled by the central franchisee, which probably contracted out the original concepts. Napkins, plates, glasses, tables, cooking equipment, point-of-purchase equipment, uniforms, etc. are all procured externally. Contracts for raw food materials, waste removal, power, and sometimes even cleaning services provide the infrastructure for the restaurant operation. The restaurant personnel are responsible for handling the customers, cooking and serving the food, and cleaning the tables (unless the customer does that, too). The developer of the franchise figures out the operational and financial model initially, and then the local owner executes to the plans. As more local operators refine the initial plans, a successful innovation can be rapidly adopted throughout the system. *The single-minded network of owner/operators sharing experience becomes a valuable asset to the successful franchise.* In this way, one successful idea (mutation in our evolutionary model) can be replicated thousands of times at distributed locations. The resulting overall organization is a mammoth, but each "cell" of the organization is focused and lean. As a result, franchised organizations are expanding in our twenty-first-century economy.

The combined techonomic power of these advantages indicate a challenging future for one-of-a-kind "mom and pop" operations in the twenty-first-century economy. As franchise operations effectively outsource more of their supporting activities, they become more economically efficient than their one-of-a-kind competitors. Less labor is required to provide the product to the consumer. In fact, many franchises go to great lengths to figure out new ways for the customer to become the labor force (pick up your own food: no waiters; clean your own table: no busboys; touch screen menu entry and automatic credit card payment: no cashiers). This is incremental Third Law techonomics, delivering the same product with less labor, by implementing inexpensive technology or process changes.

In this scenario, the Law of Diminishing Organization Size remains true if one views the individual franchise operator as the organization. The franchise operator's dependence on others for business models, inventory, designs, supplies, and even customer labor allows the franchise unit to deliver more product with less direct labor. In many successful operations, one measure of continuous improvement is the steady reduction of labor per unit delivered. Locally, firm size diminishes, while globally, the successful firm expands as more branches are added. Perhaps the best example is the McDonald's hamburger franchise, with over 31,000 restaurants worldwide and over 1.5 million employees (about 50 employees per location).

Franchises vary widely in their economic structure, but they share a similar framework of organizational success: master a local model; replicate and improve that model regionally/nationally; require local responsibility for financial success; provide strategic advantage through quantity of scale purchasing and shared experience. As the Third Law suggests, outsource the supporting functions and take ownership for the economic core of the endeavor.

SUMMARY

Three Fundamental Laws of Techonomics

Moore's Law (Law of Computational Ubiquity), Metcalf's Law (Law of the Expanding Global Network), and Coase-Downes-Mui Law (Law of Diminishing Organizational Size) are the first three laws of techonomics. The result of these laws is globalization, outsourcing, rapid product obsolescence, franchising, etc. These trends are forcefully shaping the economy of the twenty-first century.

First Law — Modified Moore's Law
(Law of Computational Ubiquity)

The cost of equivalent computing performance halves every 18 months. So, computational capabilities become economically viable for an ever-increasing number of products, and "intelligent" products become antiquated by the advance of technology in a short time.

Second Law — Modified Metcalf's Law
(Law of the Expanding Global Network)

Bob Metcalf observed that the connections of a network increase in proportion to the square of the number of nodes. The techonomic observation is that the cost of reaching anyone or finding anything on the network diminishes exponentially with the number of users.

Third Law — Coase-Downes-Mui Law
(Law of Diminishing Organizational Size)

Transaction cost analysis first developed by Ronald Coase, the make-or-buy decision, has been greatly impacted by the advent of nearly perfect information, as described by Downes and Mui. The economics of the "buy" decision are now more favorable, due to the ready access to information. The Third Law anticipates transaction costs to diminish, resulting in smaller organizations. Lean organizations tend to outsource more of their operations, thereby reducing their size. Under these competitive conditions, continuously increasing efficiency of operations becomes paramount to success.

The Franchise Effect

One successful techonomic approach to growth is franchising. In the franchise model, a central core determines strategy, methods, and processes while a distributed, financially motivated collection of owner/operators executes the model. The organization expands through replication while the individual locations improve delivery efficiencies.

Key Terms

Moore's Law	Metcalf's Law
Coase-Downes-Mui Law	Law of Ubiquitous Computing
Law of the Ubiquitous Global Network	Law of Diminishing Org. Size
outsourcing	Lean Organization
globalization	just-in-time
mass customization	Franchise Effect

QUESTIONS

1. The first three laws of techonomics are closely linked. Without the explosion of computational availability, the global network could not be accessed. Without the global network joining people and information, the access to knowledge required for diminishing transaction costs would not be available. Are these trends evident? Reversible? Inevitable? Deniable? How will the combined effect of these trends impact your organization, business, or profession?

2. An advisor of mine, Nicholas Labun (Motorola, 1993), once suggested to me as I wrote a business plan, "Think of the implications of your product as the costs are driven to 'the cost of nothingness.'" At first I drew a blank. But soon, Moore's Law brought his comments into focus. This is the trend of all digital technology: to be cheaper tomorrow than today, and ultimately, to be so inexpensive that the masses will use it ubiquitously. Pick one electronic product of personal interest and describe your observations of this product's journey to "the cost of nothingness."
3. As the twenty-first century opens, "A computer on every desk" is supplanting the economic mantra "a chicken in every pot." This new mantra is made possible by Moore's Law and made necessary by Metcalf's. The "digital divide" is a term used to describe the disparity between those who have access to the digital infrastructure (computers, web, etc.) and those who do not. The relentless march of techonomics continues to decrease the cost of entering the digital world. What are the implications for the "digital divide"? Is it an economic divide or a cultural one?
4. At first blush, the Law of Diminishing Organizational Size seems to predict that future employment will diminish. Not necessarily. As organization size diminishes, organizations will increase both in number and in specificity of purpose. We can anticipate more organizations with fewer employees per organization and with a more efficient and effective structure. Terms like "right sizing" and "lean engineering" have arisen to describe this phenomenon. Study an organization that you are familiar with (or get corporate performance numbers from annual reports) and track their annual revenues per employee over a 10-year period. The Third Law anticipates that healthy organizations would exhibit a steady increase in this metric.
5. In seeming contrast to the Law of Diminishing Organizational Size, franchises are expanding in the retail, restaurant, hospitality, wholesale, and service marketplaces. A small group perfects the operational details and then sells this knowledge to others, who establish operations in distributed locations. A much larger percentage of franchise operations succeed in comparison to the businesses of individual entrepreneurs who pursue their own venture. Why? Are there techonomic implications of the first three laws that work in favor of the franchisee?
6. According to Ray Kurzweil in *The Age of Spiritual Machines,* by 2020 there may be a $1,000 computer with the storage capacity and calculation inference capability to rival a human being. Because computer science and engineering is developing exponentially, that $1,000 computer on the desktop in 2040 may have the storage capacity and calculation inference capability of the entire living population on Earth! What are the implications of such computing to your education? Privacy? Organization? Life? What will a computer with these capacities be able to do in 2020 (or 2040) that it cannot do today?

REFERENCES

1. Wikipedia, Moore's Law, Wikipedia, The Free Encyclopedia, 2005, http://en.wikipedia.org/wiki/Moore's_law.
2. Wikipedia, Metcalfe's law, Wikipedia, The Free Encyclopedia, 2005, http://en.wikipedia.org/wiki/Metcalf%27s_Law.
3. Wikipedia, Ronald Coase, Wikipedia, The Free Encyclopedia, 2005, http://en.wikipedia.org/wiki/Ronald_Coase.
4. Downes, L. and Mui, C., *Unleashing the Killer App: Digital Strategies for Market Dominance,* revised ed., Harvard Business School Press, Boston, 1998.
5. Kurzweil, R., *The Age of Spiritual Machines,* Penguin Group, New York, 2000.
6. Coase, R.H., The nature of the firm, *Economica, New Series, 4, 386, 1937,* http://people.bu.edu/vaguirre/courses/bu332/nature_firm.pdf.
7. Friedman, T.L., *The World is Flat: A Brief History of the Twenty-First Century,* Farrar, Straus and Giroux, New York, 2005.

6 Emerging Twenty-First-Century Techonomic Business Models

INTRODUCTION

Technologies based on the first three laws of twenty-first-century techonomics empower us to conduct commerce and shape the organizations in new ways. New business models are developing that were neither conceivable nor feasible a decade ago. It is instructive to study thriving, industry-leading companies that have pioneered new models, leveraging emerging technology into lasting economic advantage. Some of these companies make and market high-tech items, while others do not, but all of them *use* technology in alignment with twenty-first-century techonomic trends to make numerous aspects of their business more effective, including:

- Minimizing inventory
- Meeting specific customer desires
- Minimizing operating cash requirements
- Increasing operating hours
- Increasing number of outlets
- Deploying capital more wisely
- Globalizing production and support
- Magnifying per employee transactions
- Streamlining and focusing marketing methods
- Eliminating traditional middlemen in the sales/distribution channel
- Eliminating retail overhead

The following sections give concrete examples of successful businesses and how they are using technology to rewrite the rules of competition. These businesses are applying the laws of twenty-first-century techonomics to their great benefit. Others must understand and apply the same laws if they are to survive in the competitive marketplace.

The philosophy of supply and demand has guided free-market economic thinking for the last two centuries, but Adam Smith developed these principles before the advent of mass production, heavier-than-air flight, computers, satellite communications, and the Internet. Humankind now has the ability to manufacture any product in a quantity that exceeds demand. We can market any product to the entire planet, and distribute any product ubiquitously, from any location. Now, deployment of resources to provide a return on investment should be determined as much by competitive positioning as market need. Without a strategic and defensible competitive advantage, gains in today's marketplace are short lived. A competitive advantage can be continuously improved by productive utilization of emerging technology in innovative products or in transforming the marketing and distribution channels. *Entirely new methods of cash management are also made possible by understanding and implementing the first three laws of twenty-first-century techonomics.*

Many successful companies are learning to manufacture only when demand arises and minimize the time lag between demand and fulfillment. The dance of supply and demand is orchestrated by wise use of technology, market positioning, and deployment of capital. Technologies are selected based on how they support the efficient production of products/services and defensible market positions based on company assets (intellectual property, distribution network, trade secrets, mind share, etc.). These defensible market positions are what Jim Collins, author of *Good to Great,* calls the "Hedgehog Principle."[1] The armored, spiky ball into which a hedgehog rolls itself when threatened affords a nice metaphor for a highly defensible market position!

As we examine the following business models, we shall use techonomic metrics to reveal reasons for success of each business. The techonomic thought process is embodied in this approach: study a company or organization and its fundamental marketplace, understand how its business model functions at its core, and create a measurable quantity that combines a technology and an economic contribution to track the company performance. Learn to apply the techonomic approach to your own endeavors, and you will become more effective in your decisions for deploying resources.

POSITIVE CASH-FLOW MANUFACTURING: DELL

Dell Computer has been a pioneer in the use of technology to revolutionize the business of the personal computer industry.[2] The Dell model gives tangible meaning to terms like mass customization, just-in-time delivery, rolling warehouses, inverted cash float, and virtual retailing. All these terms relate to business practices made possible by embracing technologies that have changed traditional business practices. Dell has rewritten the rules of PC manufacturing and distribution. The PC industry will never be able to return to its former practices, and those who hesitate to adopt the new, techonomically driven approach will slowly (or rapidly) be overtaken by the fittest companies.

The key technologies upon which Dell has based its changes are the PC itself and the Internet. Since the first two techonomic laws deal with the ubiquity of computing and the Internet, Dell is the perfect example of playing to the strengths of techonomic trends. Without the massive penetration of Internet-connected PCs available to business and home customers, the Dell approach would not be so dominant. But with the high penetration of networked PCs and the systems that Dell has developed to optimize mass customization, Dell has masterfully created a fulfillment system that is difficult for competitors to overcome. The approach has two pillars: give customers exactly what they want, and manage operational cash flow by using favorable terms on both the seller and the supplier transactions.

Dell is committed to giving customers exactly what they want as quickly as possible, one customer at a time. Mass customization. Like the Wendy's hamburger with 256 possibilities made from eight different ingredients, Dell stacks up a computer from options on currently available components (microprocessor, memory, disks, monitors, modems, operating system, etc.) to serve up a smorgasbord of product possibilities. Good products configured exactly the way you want them, when you want them, 24/7/365 thanks to the Internet.

Think how hard this would be in a "paper" world. Because computer components change so rapidly, the order would be outdated before the ink was dry on the form. The pricing would also require a spreadsheet to calculate as the combinations from different vendors at different margins were summed to determine a viable margin at a competitive price. In the traditional paper world, the best approach would be to make a lot of one configuration (quantity of scale to be competitive) and market the configuration until they were all sold. This requires warehousing and inventory costs to support the mass production run. Consider again the implications of Moore's Law. The inventory that is aging has at best an 18-month competitive life, and the manufacturer risks never selling units produced in a traditional manner, particularly if they are competing against the mass customization model.

Enter the Internet. Now the customer can enter a virtual storefront on the World Wide Web, select the exact configuration of computer components desired, place the order, and buy via a credit card. Dell gladly fulfills the order with guaranteed payment in hand, knowing what the customer wants to buy, knowing that the sale price is predetermined, and profiting from the fact that there is no additional distribution channel cost (middleman) other than a modest shipping cost. No inventory carrying costs. No distribution channel commissions. No finished product returns due to time-limited shelf life. The Internet transformed the sales approach, cycle, and channels. But it does even more related to suppliers and cash flow.

Traditional manufacturing systems obtain parts from suppliers, pay for the parts, inventory the parts until ready for manufacture, assemble the parts into a product, ship the products to retailers, and then sell products to customers generating revenues many weeks after the component parts were purchased. This approach is shown in Figure 6.1, the key elements in a manufacturing cash-flow model for a traditional manufacturer. Even though favorable terms can be negotiated, the cycle from procuring components until selling the product via retail channels is long and requires capital to fund its growth. As production increases, so must financing for production in the pipeline.

Use of capital to finance inventory was an accepted business practice until recently, when just-in-time inventory management approaches (JIT) came along. ***By applying JIT on the supply side, with mass customization direct selling on the distribution side, Dell has developed a twenty-first-century business model that actually creates an increasing positive cash flow as production increases.*** Figure 6.2 shows the key elements for the Dell positive cash-flow manufacturing model. The *Inverted Cash-Flow Model* is so named because a traditionally negative cash flow required for work in process and inventory has been inverted into a positive

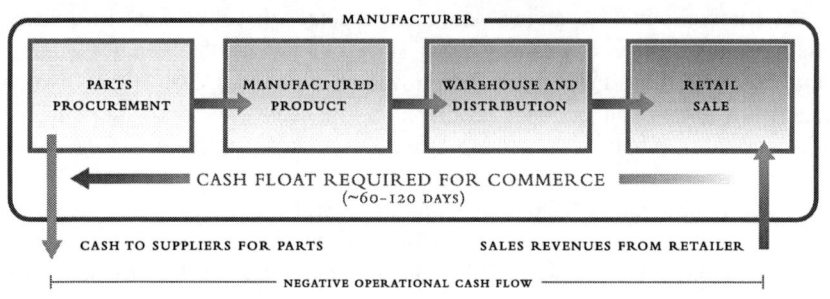

FIGURE 6.1 Key elements in a manufacturing cash-flow model for a traditional manufacturer.

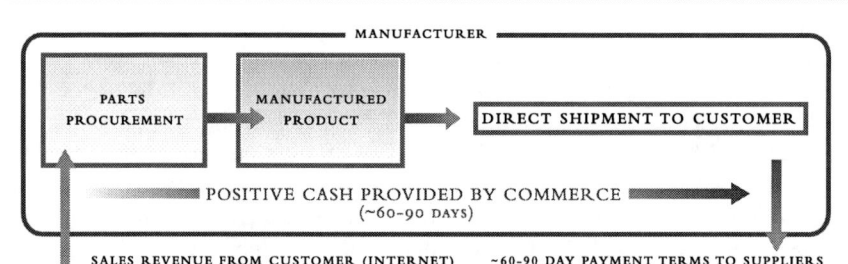

FIGURE 6.2 Key elements for the Dell positive cash-flow manufacturing model made possible by the global network.

cash flow. Naturally, Dell loves this model; they get the customer's cash (electronic payment) for the finished product (assembled computer) long before they must pay the suppliers of computer components. Electronic data interchange, World Wide Web, wire transfer of funds, rolling warehouses, just-in-time delivery, plant-to-door package delivery networks, and the wire transfer of funds are a few of the key twenty-first-century technology-backed systems that make possible the Dell model of inverted cash flow.

Compare the two figures and observe the relative location of revenues and expenses in the models. This is the most important first-order result of the application of technology to Dell's operations, but there are noteworthy second-order effects that contribute to the effectiveness of the model.

When a customer goes on the Web and orders a new PC, the entire configuration is defined. The first person to see this configuration is the Dell worker assembling the finished product. *All the communications to component suppliers, distribution centers, purchasing agents, and trucking fleets have been automatically dispatched based on the customer's order and the sequencing of production.* Component suppliers receive specifications for parts in a sequential order to match the production staging logistics. At a predetermined time, components from numerous suppliers converge at the Dell plant for assembly. Upon assembly and software loading, a burn-in period follows. The package delivery partner picks up the PC for its journey to the customer as the wire transfer of the customer's credit card is processed. This whole process often takes less than 48 hours.

Meanwhile, the component supplier provides extended terms to Dell in order to secure a large and growing business relationship. Payment terms may be anywhere from 30 to 90 days, and beyond. Since Dell has no warehouse, Dell receives cash from the customer well in advance of the payment required to the supplier. Up to three months of positive cash flow is provided for some portions of the transaction.

The financial benefits of this total approach are many, significant, and obvious: positive cash flow, no warehousing costs, no carrying inventory, no retail distribution inventory, and no product aging costs. No surprise that Dell has rocketed to the top of the list of PC suppliers. By contrast, Compaq was merged with HP because its distribution channel model failed. Even with increasing sales, price pressure from the streamlined Dell model caused layoffs at HP. IBM, the originator of the PC platform, sold its PC interest to Lenovo in China. This is an attempt to leverage the lower-cost Chinese labor and maintain a presence in the market (and to service a large and growing Chinese market).

While reducing the manufacturing labor cost content of the PC is a traditional approach that should yield results, it is the mass customization and inverted cash-flow model that is the strength of Dell's continuing competitive advantage. Dell's suppliers have become their bankers and inventory managers. Others will emulate the techonomic strategies of companies like Dell or become its victims.

Free cash flow is the cash remaining after all expenses (net income plus amortization and depreciation minus operating expenses, capital expenditures, and dividends) including investments have been paid. Free cash flow differs from earnings in that it accounts for capital expenses as it occurs rather than depreciating it over many years. Free cash flow is meant to capture all real cash outlays of the present. Inverted cash flow describes a different concept altogether. It fundamentally describes and quantifies a process whereby an organization receives payments for its endeavors before it pays its suppliers. In traditional cash flow, money moves out of a company to pay for supplies before revenues return to the company to pay for the end product. Electronic commerce and savvy negotiation has allowed the tables to be reversed by some key organizations, hence the development of a new techonomic financial term, *inverted cash flow*.

Let us say you want to consider using the inverted cash-flow model as part of your overall techonomic strategy. How can it be measured and improved? A very simple techonomic metric can be used to measure cash float for an endeavor. We will call it **Endeavor Cash Float (ECF), the combination of time and cash flow contained between the payment for and the costs incurred in the process of delivering an endeavor.** The key to the ECF techonomic metric is having the systems in place to measure the components that constitute it. ECF provides a measure for the cash flow produced or consumed from the manufacturing through the sales cycle. Shorten the sales cycle, and the metric increases. Negotiate better terms with your suppliers, and the metric increases. Reduce the duration for parts inventory, and the metric increases. This metric can pinpoint where to make critical improvements to ECF. Carefully note that the ECF metric does not indicate whether a given product sale is profitable (it does not consider the cost of labor, cost of marketing, etc.). It simply quantifies the time value of a product's material cash flow as the product moves through the procurement, production, and distribution pipeline from supplier to customer payment. Increasing the ECF reduces, or can eliminate, the need for capital to fund cost of goods.

The ECF sums the difference between sales price and cost of each component times the number of days of float — that is, the number of days between receiving customer payment and having to pay the supplier. The days of float can be positive or negative. If customer payment is immediate and all goods are JIT delivered with 30-day terms, then the number of float days is 30, and the ECF is the sales price minus the cost of goods in dollars times 30 days. If you are a startup manufacturer with no supplier terms, and your channel is retail with 90-day payments, your float is –90 days times your sales price minus the cost of goods (ouch!). Here is the Endeavor Cash Float techonomic metric.

ENDEAVOR CASH FLOAT TM

Assumptions:

1. Purpose is to monitor product production/sales cycle cash performance.
2. Goal is to maximize positive $ × Days.

$$TM_{\text{Product Cash Float}} = \Sigma\, ((\text{Price} - \text{COGS})_i \times (\text{FLOAT}_i))$$

where
Price = price for product ($)
COGS = cost of goods sold ($)
FLOAT = days between customer payment receipt and payment of supplier (days)
i   = index of components

Since multiple vendors under unique terms often provide components for a product, the ECF computation may require some data mining for a given endeavor. Application of the Pareto Principle with the EFC computation will provide a systematic approach to releasing cash from work in progress. **Alfredo Pareto's 80–20 rule states, in essence, that 80% of the benefit can be obtained from 20% of the contributors.**[3] In basic terms: solve important problems first to get maximum benefit from your efforts.

In summary, the positive manufacturing cash-flow model seeks to receive revenue from the customer before paying the supplier. The greater the metric becomes, the more time-efficient the production/distribution cycle. Technology can be leveraged to create logistics supporting just-in-time delivery, customer-direct procurement, automated billing, material processing, and mass customization. The resulting system eliminates retail inventory costs, minimizes work in progress, reduces product obsolescence, and creates opportunities for positive cash float between the customer electronic payment and the compensation of suppliers. *The ultimate example of ECF would be total outsourcing of finished products without having to pay for them until after they were sold* — our next example!

POSITIVE CASH-FLOW RETAIL DISTRIBUTION: WAL-MART

Like Dell, Wal-Mart has been a judicious user of technology. This has been a significant key to its market dominance. While the Third Law of Techonomics predicts that organization size will shrink as transaction costs reduce, Wal-Mart has grown rapidly by any measurement over the last 20 years.[4] This seems to defy the Third Law, unless you consider the means of expansion: franchise. Wal-Mart found a model that worked and then replicated it massively, all the while putting in place systems that allowed smaller work forces to accomplish more. Wal-Mart is constantly increasing the productivity of its operations via continuous improvements, many based on technological innovations.

Using twenty-first-century data management systems, Wal-Mart places the inventory control responsibility for their shelves into the hands of their suppliers. Suppliers are allocated shelf space based on Wal-Mart's projections. Suppliers are responsible for keeping that shelf space full of viable products, and they carry the cost of this inventory, often for months after the product is sold. Suppliers are left with the responsibility for unsold merchandise at season's end if they happened to overstock Wal-Mart. The key to this system's success is an electronic data network linking vendors with inventory information from Wal-Mart central distribution centers, retail store shelves, and store cash register transactions. Ideally, vendors have "perfect information" in terms of their company's collaboration with Wal-Mart (but they do not get information about Wal-Mart's other vendors; that is shielded). They always know how many of their own products are on what shelves at what stores and how fast they are selling each day. *It is the supplier's responsibility to have the right product quantity available at the distribution centers to meet the needs of the store network.*

In this operating model, Wal-Mart looks like a giant consignment shop. Vendors place products on borrowed shelf space and are paid for only those products that sell, with payment terms that are favorable to Wal-Mart. It is a very challenging environment for vendors, but the volume of the opportunity is also very enticing. The story of one man who said no to Wal-Mart, Jim Wier, then the CEO of Simplicity, makes interesting points about how the specter of Wal-Mart changes the economics of competition, even if a company chooses to say no to being a supplier to them.[5] Wal-Mart minimizes risk by giving inventory responsibility to the vendor. They also maximize positive cash float by negotiating payment terms that extend beyond the period of the anticipated cash sale to the customer.

Here is a simple example of the process to illuminate the model. Suppose a publisher is seeking to sell a book through Wal-Mart. The publisher approaches Wal-Mart with a proposal to garner the coveted shelf space. After some challenging negotiations, Wal-Mart agrees to take 10,000 books for 90 days, with an additional 90-day payment term. The books sell for $30 and the publisher is to receive $20 of that price for all books sold during the 90-day sales period. Everybody is happy. The publisher now has a potential $200,000 book-selling contract with Wal-Mart.

On day one, the publisher ships 10,000 books to the Wal-Mart distribution center. They sell well and on day 89, Wal-Mart collects all remaining stock from all stores (say 1,000 books) and ships them back to the publisher. No payment yet; the terms were 90-days after the sales period. On day 91, Wal-Mart reorders another 10,000 books to begin selling again, and the cycle repeats. Still no cash in hand for the publisher, who is now out printing costs for 20,000 books and has 1,000 of them on hand with Wal-Mart stickers on them! They must be sold for ten cents on the dollar to a "seconds" dealer. Day 180 comes, and the publisher receives a check for $180,000 ($20 x 9,000 sold in the first three months). By now, Wal-Mart has sold the second 9,000 (returning again 1,000 not sold) and has received cash revenue of $540,000 with a gross profit (profit before operating and overhead costs are deducted) of $180,000 (18,000 x $10 per book sold). This cash management discipline eliminates all inventory risk at the expense of a small increase in transportation logistics. The Wal-Mart logistics system is already in place though, so the incremental cost of shipping the extra product back to the publisher is minimal.

The gross markup of about 35% is typical of a retailing industry, but Wal-Mart's positive cash float results from a system of data management (technology), automated distribution (technology), and savvy contract negotiation (economics) that fuels the growth of a highly successful twenty-first-century enterprise. Meanwhile, Wal-Mart's suppliers and their competitors both languish while they try to understand the dynamics of this arrangement. Like them or not, Wal-Mart must be admired for their systematic approach to maximum efficiency, passing on many cost savings to the customer and thereby making their competitive position even more substantial.

The success flywheel moves faster and faster, and the inertia of the 1,000 things Wal-Mart did to get into a dominant position creates a crushing push against competitor entry. Massive buying power allows negotiation of favorable terms, both cost and payment cycle. The rolling warehouse network, backed by the automated distribution system, reduces inventory costs and puts the right products on the shelves in the right season. The combination of these advantages has made Wal-Mart a very tough competitor in many sectors. The company systematically enters new markets with a goal to obtain a market position in excess of 30% of the entire market segment, placing it in the top two retailers in any given category.

Rank upon rank of companies have been transformed or displaced by the Wal-Mart juggernaut. When Wal-Mart enters a small U.S. community, the first to feel the crush of competition are the "mom and pop" establishments. With limited buying power and operating hours, they are no match for the diversity of goods or the competitive pricing of Wal-Mart. Most of these small operations last no more than two to five years on goodwill, perseverance, and savings. The second rank to fall is direct competitors. Sears, K-Mart, and Service Merchandise have either ceased business or faded to mere shadows of their former operations. Target and Costco are the remaining direct competitors. Now Wal-Mart is aggressively entering the grocery market. Winn-Dixie has recently declared bankruptcy, and long-standing chains like Kroger and Albertsons are battling for customer loyalty.

Wal-Mart's success is now causing the competitive restructuring of major product manufacturers. Proctor and Gamble has merged with Gillette, combining two companies that had previously been formidable competitors. The first reason listed in information to shareholders was the necessity of putting the combined corporation into a better bargaining position with Wal-Mart. The recently hired CEO of Sara Lee said her company would return to making baked goods, where there was a reasonable margin, and sell off its clothing lines, because there was no margin remaining in clothes sold through Wal-Mart's channels, which had accounted for over one third of their clothing sales.

The coming impact of Wal-Mart's dominance may well shift the international trade balance. For some time, Wal-Mart's buyers have been commissioned to find the right product at the best price anywhere in the world. Global sourcing has been the techonomically logical direction for filling its huge distribution channel. Low-cost Chinese labor, guided by focused product specifications from Wal-Mart and interconnected by the global digital network, has caused a massive shift in global consumer goods manufacturing over the past decade (from Japan and the U.S. to China). International product collaboration, manufacturing inventory management, and logistics (technology) have combined with the timeless natural selector of lowest-cost labor pool (economy) to drive these production trends. Figure 6.3 shows the shift in U.S. imports from various key trading partners over the last 20 years.[6]

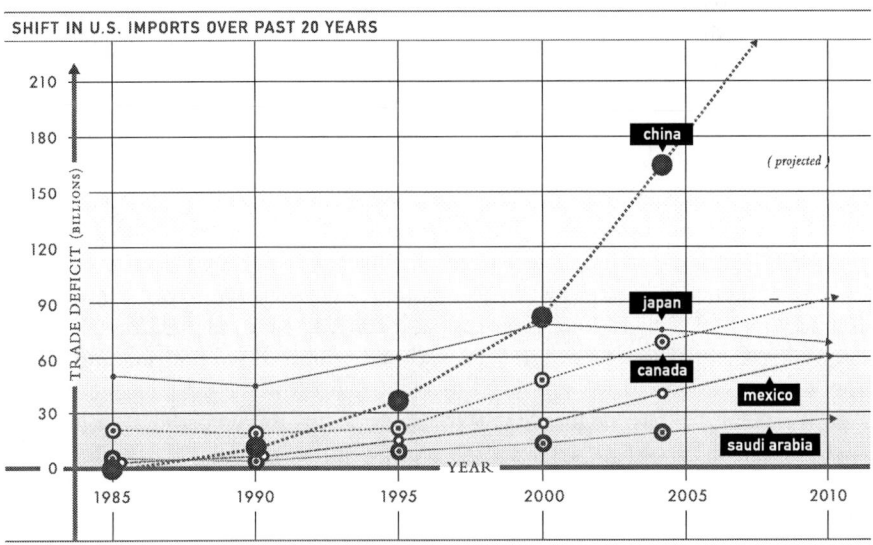

FIGURE 6.3 Shift in U.S. balance of trade (deficit) from various key trading partners over the last 20 years.

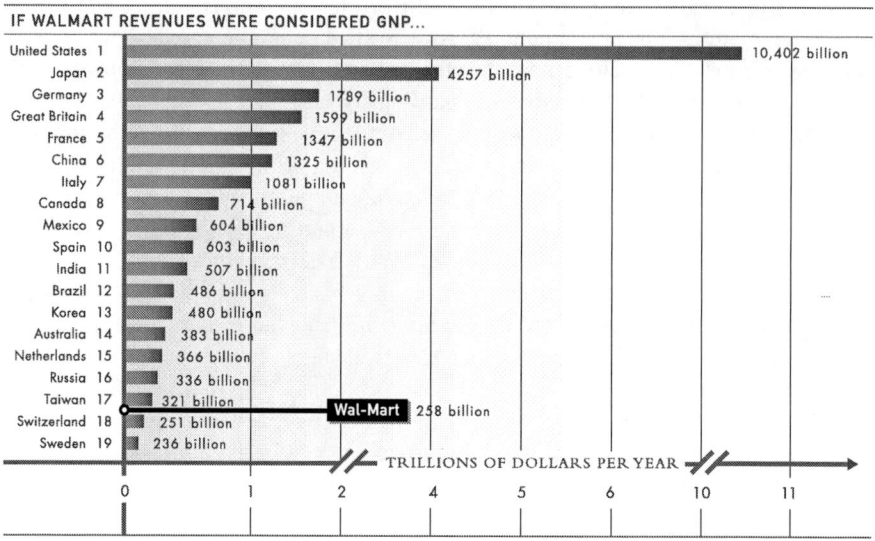

FIGURE 6.4 Ranking of Wal-Mart as a country if its revenues were considered a gross national product (2003 data).

Wal-Mart is nearing "nation" status in its size, negotiating ability, and influence in the world economy. Figure 6.4 shows the ranking of Wal-Mart as a country if its revenues were considered a gross national product (in 2003). *If Wal-Mart were a country its "revenues as GNP" would place it as the 18th largest economy in the world*, right behind Taiwan and in front of Switzerland.

Even more impressive from a trend standpoint, Wal-Mart and China are the only two on the list growing at double-digit rates. Within 5 years, China would project to have the world's third largest GNP, and Wal-Mart would be nearing the top twelve (if it were a country). Wal-Mart has almost single-handedly created a massive distribution channel for Chinese goods in the U.S. over the past decade. The near-term result for U.S. consumers has been a slowing of inflation as consumer goods have remained inexpensive. The long-term result may be that competition for the world's basic natural resources (energy sources like petroleum, building supplies like steel, and raw materials like asphalt) will increase, causing prices to escalate.

The techonomic metric that measures the distance between Wal-Mart and its competitors is the direct item price comparison. Wal-Mart has traditionally positioned itself as the low-price provider, satisfied with a 35% retail markup on the goods it sells. If it can drive a provider to a lower cost with the same payment terms, then Wal-Mart will pass on the savings to its customers. Its competitors, with a more costly distribution system, less favorable payment terms with suppliers, and less purchasing power to drive down supplier costs, are left to price their offerings as best they can while still making a profit.

The techonomic metric for competition margin measures the performance difference between Wal-Mart and its challengers in any market. This metric relates the shelf price for a product from a competitor to the Wal-Mart supplier cost and retailing cost for the same product. Retailing markup is traditionally 35%, which includes the retailer's selling costs. As the retailing and supplier costs are reduced in the Wal-Mart system, the competitive margin increases. If the metric is less than one (1.00), Wal-Mart loses money on the transaction. An increasing value for the metric beyond 1.35 reveals opportunities for pricing approaches that undercut vulnerable competition.

COMPETITIVE MARGIN TM

Assumption:

1. Purpose is to monitor item-by-item pricing advantages as a result of operational effectiveness.

$$TM_{Competitive\ Margin} = \Sigma\ (PRICE_{Competitor}/(COST_{Supplier} + COST_{Retailing}))$$

where
$PRICE_{Competitor}$ = competitor selling price of product
$COST_{Supplier}$ = cost of product from the supplier
$COST_{Retailing}$ = total overhead cost assigned to product including facilities, distribution, transport, etc.

Wal-Mart has been a leader in adopting technology that maximizes the competitive margin. Since they have no control over competitor prices, there are only two contributors that Wal-Mart can affect to increase this metric: supplier cost and retailing cost. Using global sourcing and electronic inventory management, while negotiating favorable terms, Wal-Mart continually drives down their cost of goods. Using efficient warehousing and distribution channels and technologies that increase productivity, Wal-Mart reduces the cost of retailing.

Wal-Mart regularly rotates their buyers, those employees who procure products from suppliers. In doing so, the company assures that fresh eyes are continuously focused on the goal of the company and not clouded by comfortable relationships. Wal-Mart regularly "shops" its products to new potential suppliers to see if a lower-cost supplier can be identified, locally or globally. Bids are formally requested and used as bargaining tools to extract lower prices from current suppliers. While these practices are challenging, the result is a supplier network with clear expectations and no chance to rest on past performance.

Wal-Mart was an early adopter of barcodes for inventory tracking (and is now a first mover for radio-frequency identification tags, RFID, to further automate stocking processes). Their electronic data for suppliers are transparently integrated, passing the responsibility and accountability for product availability to the suppliers without a middleman. This approach is similar to the one used by Dell to minimize inventories.

Wal-Mart's use of technology and economic strategies has made it a powerful leader with a strong competitive position. So powerful, in fact, that it is a silent force behind the modification of eminent domain laws in the U.S. In the past, eminent domain laws kept the government from seizing private property for any reason other than the public good (highways, roads, municipal buildings, utility lines, etc.). Recently, these laws have been under pressure from developers desiring to condemn property for retail facilities that offer the promise of generating substantial tax revenues. The 2005 U.S. Supreme Court case of *Kelo v. City of New London* is an example of the broadening of eminent domain to accommodate one private use over another, diminishing the rights of landowners.[7] While this case did not involve Wal-Mart in any manner, the ability of commercial developers to use the government as a means to seize land for economic development purposes was greatly strengthened. **Techonomics predicts that many more eminent domain disputes based on economic developments will arise in the future as those in power act for what they deem to be the public good — increasing the tax base.**

In summary, Wal-Mart uses a highly automated and continuously monitored inventory system to place the responsibility for product availability in the hands of its suppliers. The suppliers, virtually eliminating Wal-Mart's financial risk, commonly support the carrying costs for products on Wal-Mart's shelves. By using digital networks to tighten the linkage between supplier, retailer, and customer, while simultaneously negotiating favorable payment terms with suppliers, Wal-Mart generates a positive cash flow as its revenues grow. Wise use of other people's money via optimized data management of inventory logistics is a repeating techonomic theme of successful twenty-first-century businesses.

DEBTLESS FACILITY EXPANSION: WALGREENS

The drug store retail chain Walgreens provides another example of focused use of technology and creative access to other people's money to rapidly grow a very fit organization. Walgreens' simple business vision is to become the McDonald's of pharmaceuticals.[8] This clear parallel connotes clean, efficient operations and a large number of outlets at prime locations throughout the country. Walgreens has pursued this goal using a masterful techonomic approach, combining technology for effective customer service with the economics of other people's money to finance facilities without incurring debt. This approach may not work exactly the same way for everyone, but it exemplifies the foundation of techonomics: thinking that takes advantage of technology and creates financial opportunities by leveraging resources.

Walgreens has been an early adopter of technology to manage customer records. In the 1980s, the company created its own satellite communication system to link its stores. The network allowed a customer of any Walgreens location to get refills or other help related to prescriptions at any other store on the network. With the advent of managed healthcare and prescription coverage for medications, Walgreens was positioned to manage insurance billing for customers, making their transactions easier.

Collecting store information from many outlets allows the collective whole to learn from the localized patterns of operations. Over the years, Walgreens has been able to use demographic information to target expansion opportunities and then use individual store performance to determine when full 24-hour operations are financially justified based on revenues from different profit centers within the store. This detailed knowledge of financial performances and variances is a valuable component for predicting the performance of potential store locations. As the number of distinct locations grows, the amount of store performance data increases, as does the customer database (population density, income, age, medical expenditures, etc.). The successful local operational model can be replicated hundreds of times in communities across the nation.

Confidence in the long-term success of individual operations provides a bargaining position for establishing partnerships. Knowledge is power. Walgreens knows the size of the store and the site needed. They know the location demographics that maximize financial return. In a given community, they know where they want to be located, down to which side of the street is best! Now all they have to do is get a store open and execute their business.

How does Walgreens rapidly build a large number of stores without incurring a burdensome debt or equity dilution? Like Dell and Wal-Mart, they find a unique way to use other people's money. Walgreens' source is the local real estate developer. Via a contractual partnership that affords the developer strong and lasting financial returns, Walgreens induces the developer to finance the storefront.

In brief, the deal goes like this: Walgreens finds a choice location and approaches a few developers in the area, seeking a partner. Walgreens states its willingness to enter a long-term lease for the store at a premium rate (say, $25 per square foot for 30 years). Compare this opportunity to a one- or two-year lease from less stable occupants at $18 per square foot, and you see why many developers are willing to fund the Walgreens facility. With a known return on investment, the building and occupancy risks are minimal for developers, and they are eager for the opportunity. Walgreens knows, within a small variance, what revenue per square foot a store will produce and is creating a successful proposition. Both parties win. Walgreens expands without incurring significant debt.

A techonomic metric that can be helpful in the determining store performance and product merchandising is the customer value metric. Measuring the cash received as a result of each customer visit, this metric provides a means to track the productivity of the store footprint, the product offerings, and the anticipated revenue based on store traffic.

CUSTOMER VALUE TM

Assumptions:

1. Purpose is to monitor store performance as retailing approaches are changed including merchandising, product mix, floor plan, advertising, etc.
2. The dependability of this measurement improves as the data is compiled from more locations or from a single location over a large number of customer visits.

$$TM_{\text{Customer Value}} = CASH_{\text{Customer}}/VISIT$$

where
  $\text{CASH}_{\text{Customer}}$ = dollars sold to a customer
  VISIT = one customer visit

PREDICTABLE ANTIQUATION: INTEL

Planned obsolescence has been an engineering approach to design for as long as there has been mass production. Nothing lasts forever. But, how long should something be expected to last? At its best, the process of obsolescence design (or predicting failure modes) is the combination of economic and technical considerations. Not only is it a question of how long something should last, but a question of the cost for increasing the operating life of a product. Such decisions on mechanical parts, like bearings in an automobile, determine the useful life expectancy of the product.

For mechanical systems, as design margins and safety factors diminished with the advent of computer-aided design, the expected useful life of products also decreased. Washing machines and vacuum cleaners of earlier times often served a household for 20 years or more. Today's offerings work well, but have been "unitized" (designed with nonserviceable components) and designed with fewer margins. Result: more rapid obsolesce. Businesses in the old economy (mechanical/industrial) developed their product obsolescence around failure modes for the product (bearings, gears, belts, etc.). Businesses in the new economy (electronic/information) develop their product obsolescence cycle around the antiquation of their products due to advancing capabilities (speed, features, compatibility, tax law changes, etc.).

Moore's Law has lead to a completely new form of planned obsolesce, something I call predictable antiquation. ***Predictable antiquation estimates when product abandonment will occur due to technological advance, not due to product failure.*** Electronic products quickly fall into disfavor, often before they fail to function. Antiquation may result from computational speed (PCs), operating system incompatibility (software products), image resolution (digital cameras), storage capacity (MP3 players), network incompatibility (modems, cell phones), or system integration usurpation (personal digital assistants). Frequently, it is a battery that will no longer hold a charge that is the last straw for a frustrated consumer.

Decreasing product life cycles and a continued battle to maintain market position in the electronic marketplace have been the direct result of the rapid advances in electronic technology. The emerging and overwhelming phenomenon of product abandonment, rather than product failure, has given rise to business models that succeed by significantly surpassing previous product performance capabilities, inducing a shift in consumer buying patterns. Companies cannibalize themselves (i.e., they create new and better offerings while they still hold a leadership position in the marketplace with their previous offering) to feed the consumers' hunger for more, faster, better, cheaper. Intel Corp., developer and manufacturer of microprocessors, has been the leader in pursuing a techonomic business model based on predicted antiquation.[9]

Gordon Moore was a cofounder of Intel and also the originator of Moore's Law. Predictable obsolescence was the business DNA upon which Intel was founded, and it was based upon Moore's technical observations. Basically, if the company ever stood still for two years with an existing product, it would be surpassed by a wave of competitors continually improving their offerings. From the start, Intel organized their design process, marketing launches, and financial planning around the concept that they must be the first to innovate the next generation of product.

The technological advances revealed in processor performance gains were remarkable and positioned Intel as the de facto leader in the microprocessor marketplace. The financial commitment in research, development, marketing, and fabrication was justifiable only with the technical understanding of this model in mind. Before Intel, it would have been considered ludicrous to introduce a product that would usurp market share from your own product when you already controlled most of the market. Today, market leaders in consumer electronics have to execute their product plans in this way simply to remain viable in the marketplace. New features, more memory, more colorful display, greater compatibility, etc. are the requirements of an ever-more-informed consumer. The instant information and procurement paths offered by the Internet do not let producers hide behind brand name or strong distribution channels.

Another strategy Intel used for staying on top of predictable antiquation was research into new applications that would demand greater product performance — research into how innovators (small companies and academics, for example) were using Intel products in demanding ways or with new peripheral devices to stretch the limits of performance. The PC continued to improve as application demands continued to expand. Color, sound, networking, display resolution, print resolution, digital cameras, digital video, modems, broadband, graphics, animation — the list of software and peripheral advancements related to the PC is lengthy. Most of these advances occurred in laboratories or engineer's garages a few years before the speed of the PC made their performance possible or economically feasible. But each one placed new demands on the performance of the PC, from speed to memory to compatibility. The new advances also created new markets with expanding demands for PCs. PCs moved from the workplace to the home, and from the desktop to the laptop. Intel's wisdom, in addition to predictable obsolescence, was the encouragement and creation of applications that accelerated the demand for better performance.

The increasing performance of the microprocessor has not come without a price. One major contributor to improvements in operating speed has been manufacturing methods that pack more components into a given area of silicon. As designs shrank, manufacturing tolerances have become very demanding, leveraging from numerous improvements in supporting technologies including material purity, clean-room techniques, process control, etc. New fabrication facilities and advanced equipment are required for each new generation of microprocessor. Through technology, microprocessor improvements will continue to track Moore's Law expectations for the foreseeable future.

Economically, is there an end in sight for the expenditure of capital needed to build fabrication facilities for production of next-generation chips? Certainly Intel remains committed to this approach. At what point would a techonomic analysis of the market, its pricing structure, and potential reveal that the risk of taking the next step exceeds anticipated rewards? Due to the short product life cycle, about two years, the facility capital cost becomes a significant contributor to the overall consideration of product per unit cost. Below is a techonomic metric for products that have a limited shelf life and require a significant capital investment.

PREDICTABLE OBSOLESCENCE TM

Assumptions:

1. Purpose is to determine the range of viability for capital investment in the production of a specific product that has a limited market life.
2. Estimation of the volume sold is based on previous market share and current growth trends in the market.
3. Once product is out of favor, its margins and volumes diminish, limiting potential return of plant capital.
4. If the TM is below one (1.0), the endeavor may have considerable financial risk.

$$TM_{\text{Predictable Obsolesce}} = [\text{VOLUME} \times (\text{PRICE} - \text{COGS})]/\text{CAPITAL}_{\text{Facility}}$$

where
- VOLUME = Total number of product sold before end of obsolescence period
- PRICE = Average price per unit
- COGS = Cost of goods sold excluding capital costs
- $\text{CAPITAL}_{\text{Facility}}$ = Cost to design and build fabrication facility

Let us consider a rough order of magnitude approximation for the TM of a recent Intel microprocessor, the Pentium 4 Prescott. PC sales of all makes and models in 2004 were 180 million worldwide. Assume a 10% Prescott market penetration and a 2-year penetration window to estimate 36 million chips to return the production plant capital. Published product prices on the Web range from $250 to $1000, so assume lower-volume prices in the first two years, which make up the majority of sales — estimate $200 per product. Material and labor costs of goods are small relative to the facilities costs due to automation and massive production numbers — estimate $10 per chip. These values can be combined with the minimum break-even point for the techonomic metric (1.0) to determine the maximum investment that should be considered for producing the product. The resulting maximum value for a capital outlay for a production facility to return the capital would be about $7 billion. Obviously, Intel can still build fabrication plants at the cost of $1 billion and expect to make a return on investment as long as their market share remains strong. It is important to know how to sequence the deployment of new chips to the market so that a premium price is obtained for new processor introductions.

Intel's techonomic approach, successfully executed for almost 30 years, uses predictable obsolescence to stimulate demand and maintain a leadership position over its competitors. By regularly and periodically offering a better-performing product to compete with its own aging products, a business model based on predictable obsolescence can support a sustained leadership position in a rapidly changing marketplace.

BUSINESS AT THE SPEED OF LIGHT: MICROSOFT

Intel's position has also been supported by a software operating system from Microsoft. Microsoft can be viewed from the predictable obsolescence vantage point also. With each new release of the Microsoft operating system, an entirely new demand for additional software applications results due to compatibility considerations. Microsoft, by creating ever-more-capable and complex operating systems, places an increasing demand on the hardware platform's calculation and memory capacity for proper operation. Purchase of a new PC requires purchase of a new operating system — and often core applications. The predictable obsolescence cycle becomes self-fulfilling between the larger software systems requiring larger hardware that requires an operating system and new applications in a continuing spiral.

The advantage to Microsoft's predictable obsolescence business model relative to Intel's is that Microsoft's does not require large startup capital for manufacturing. The key to Microsoft's long-term success is the maintenance of barriers to entry for the PC operating system. Microsoft's approach to maintaining these barriers has been a masterful combination of embracing new technologies (extending the operating system capabilities with each release), bundling their software with numerous hardware partners, and acquiring competitors before they become threats.[10]

Microsoft has met the "open source" challenge by creating freeware versions that are not universally compatible, creating a modern-day Tower of Babble. "Open source" refers free or inexpensive software that provides an open, user-modifiable platform like Java for the Internet or Linux for computers. *Without a corporate economic beneficiary to serve as "keeper of the jewels," all free and openly extensible software is very corruptible.* Microsoft has to continually put tremendous resources into Windows to protect this system from hackers and software viruses. Ease of use, reliability, and compatibility are valuable, and Microsoft knows the customer is willing to pay for these product qualities. When browsers became a key to PC communication, Microsoft moved powerfully to create Explorer and then bundle it with the operating system, again defeating a significant competitive threat. Strong market position, with the ability to control the next system's features and release schedules, continues to propel Microsoft.

Whereas Intel has had major competitive challenges from Advanced Micro Devices, which builds close copies of Intel microprocessors, Microsoft has not been anywhere near as sensitive to competitive pressures from direct emulation of its software. Although almost any piece of technology can be copied with time and resources, it is much more difficult to copy economic market dominance that is strengthened by bundled partnerships and thousands of supporting applications.

The principal danger to Microsoft's strong business position lies in the "climate" of the cultures and economies where it seeks to grow. Since its product, software, can be illegally copied and distributed ("pirated"), Microsoft can thrive only where copyright laws are honored or enforced. As we discussed in Chapter 1, the laws of techonomics function fully only in environments that allow and respect their operation.

VIRTUAL RETAIL: AMAZON

In the *Amazon, Inc. 2004 Annual Shareholders Report,* founder and CEO Jeff Bezos provides a textbook explanation of "free cash flow" to enlighten shareholders about his company's financial strategy.[11] Amazon has done with online retailing what Dell and Wal-Mart have done in terms of structuring business operations to optimize use of vendors' cash float. *By bringing buyer and seller together in a virtual store, collecting the buyer's funds when the order is placed, and maintaining advantageous payment terms with suppliers, Amazon is able to grow its free cash flow as revenues increase.* The technology of the Internet and its availability in the homes and offices of target buyers combine with electronic transactions for ordering, billing, and collection to revamp a traditional and stable business: the bookstore.

Small, brick-and-mortar bookstores cannot measure financial performance based on free cash flow unless they have the buying power to demand product on consignment from the publisher. As the product ages on the shelf and payment terms are reached, capital is required to stock the shelves with product (books). The sales cycle is unpredictable. Compare this to the Amazon model: the product is described at a user-friendly Web site, a few pages are digitized for the customer to "flip" through, reviews are generated by the customer, and only a minimum of capital is expended for the computer system that delivers the information worldwide. Once the critical systems are in place for serving the customer base, there is only a minimal incremental cost associated with increasing the product offering by 10, 100, 1000, or 10,000 titles. Inventory turns are maximized and capital requirements minimized by keeping on hand only the most popular titles while developing a supply network that requires the seller network to maintain product until it is purchased.

CEO Bezos states it clearly: "Our most important financial measure is free cash flow per share." This is Amazon's techonomic metric constructed from financial terms. The free cash flow is enabled by wise use of technology. Additional capital expenditures can be analyzed according to how they will impact free cash flow. Free cash flow includes consideration for inventory turnover, payment terms from customers, payment terms to suppliers, and revenue growth in addition to other, less dominant factors.

FREE CASH FLOW TM

Assumptions:

1. Purpose is organizational effectiveness in generating free cash flow from operations.
2. All advancements in productivity and efficiency of operations should ultimately affect this metric.

$$TM_{\text{Free Cash Flow}} = [CASHFLOW_{\text{operating}} - REINVESTMENT\ CAPITAL]/SHARE$$

where
- CASHFLOW$_{operating}$ = Cash flow from operations
- REINVESTMENT CAPITAL = Capital support for production
- SHARE = Number of shares of stock outstanding

The concept of free cash flow is so fundamental to how successful twenty-first-century companies are growing that Jeff Bezos devoted several pages in the Amazon 2004 shareholders report to explain the concept. In the first example, note the earnings from operations are strong, while the cash flows are negative. Bezos' point is company performance goes beyond the earnings statement. The following is a hypothetical example presented by Jeff Bezos in the Amazon 2004 shareholders report as he illuminates the difference between earnings and cash flows from operations.[12]

> Though some may find it counterintuitive, a company can actually impair shareholder value in certain circumstances by growing earnings. This happens when the capital investments required for growth exceed the present value of the cash flow derived from those investments.
>
> To illustrate with a hypothetical and very simplified example, imagine that an entrepreneur invents a machine that can quickly transport people from one location to another. The machine is expensive — $160 million with an annual capacity of 100,000 passenger trips and a four-year useful life. Each trip sells for $1000 and requires $450 in cost of goods for energy and materials and $50 in labor and other costs.
>
> Continue to imagine that business is booming, with 100,000 trips in Year 1, completely and perfectly utilizing the capacity of one machine. This leads to earnings of $10 million after deducting operating expenses including depreciation — a 10% net margin. The company's primary focus is on earnings; so based on initial results the entrepreneur decides to invest more capital to fuel sales and earnings growth, adding additional machines in Years 2 through 4.
>
> Here are the income statements for the first four years of business:

	Earnings (thousands)			
	Year 1	Year 2	Year 3	Year 4
Sales	$100,000	$200,000	$400,000	$800,000
Units sold	100	200	400	800
Growth	N/A	100%	100%	100%
Gross profit	55,000	110,000	220,000	440,000
Gross margin	55%	55%	55%	55%
Depreciation	40,000	80,000	160,000	320,000
Labor & other costs	5,000	10,000	20,000	40,000
Earnings	$10,000	$20,000	$40,000	$80,000
Margin	10%	10%	10%	10%
Growth	N/A	100%	100%	100%

It is impressive: 100% compound earnings growth and $150 million of cumulative earnings. Investors considering only the above income statement would be delighted.

However, looking at cash flows tells a different story. Over the same four years, the transportation business generates cumulative negative free cash flow of $530 million.

	Cash Flows (thousands)			
	Year 1	Year 2	Year 3	Year 4
Earnings	$10,000	$20,000	$40,000	$80,000
Depreciation	40,000	80,000	160,000	320,000
Working capital				
Oper. Cash Flow	50,000	100,000	200,000	400,000
Cap. Expenditures	160,000	160,000	160,000	160,000
Free Cash Flow	$(110,000)	$(60,000)	$(120,000)	$(240,000)

There are of course other business models where earnings more closely approximate cash flows. But as our transportation example illustrates, one cannot assess the creation or destruction of shareholder value with certainty by looking at the income statement alone.

Notice, too, that a focus on EBITDA — Earnings Before Interest, Taxes, Depreciation and Amortization — would lead to the same faulty conclusion about the health of the business. Sequential annual EBITDA would have been $50, $100, $200 and $400 million — 100% growth for three straight years. But without taking into account the $1.28 billion in capital expenditures necessary to generate this "cash flow," we are getting only part of the story — EBITDA is not cash flow.

What if we modified the growth rates and, correspondingly, capital expenditures for machinery — would the cash flows have deteriorated or improved?

	(In thousands)		
Year 2, 3, 4 Sales and Earnings Growth Rates	Number of Machines in Year 4	Year 1–4 Cumulative Earnings	Year 1–4 Cumulative Free Cash
0%, 0%, 0%	1	$ 40,000	$ 40,000
100%, 50%, 33%	4	$100,000	$(140,000)
100%, 100%, 100%	8	$150,000	$(530,000)

Paradoxically, from a cash-flow perspective, the slower this business grows the better off it is. Once the initial capital outlay has been made for the first machine, the ideal growth trajectory is to scale to 100% of capacity quickly, then stop growing. However, even with only one piece of machinery, the gross cumulative cash flow does not surpass the initial machine cost until Year 4, and the net present value of this stream of cash flows (using 12% cost of capital) is still negative.

Unfortunately, our transportation business is fundamentally flawed. There is no growth rate at which it makes sense to invest initial or subsequent capital to operate the business. In fact, our example is so simple and clear as to be obvious.

Bezos' financial example reveals the importance of focusing on free cash flow as the business driver and the guide to technology adoption. In the last decade, improving free cash flow has been a defining strategy of highly successful operations. Dell, Wal-Mart, and Amazon are examples of business models that optimize performance by using technology to maximize free cash flow. Even Microsoft, by minimizing capital required for product manufacturing, is an example of creating free cash flow without needing large capital infusion for inventory.

Tomorrow's emerging opportunities will result from turning growing data networks into generators of free cash flow. If you are a corporate buyer, do not bargain for merely a good price. Bargain also for favorable payment terms within the larger context of business cash flow. If you are on the wrong side of free cash flow, your company's capital is being usurped to grow someone else's bottom line.

VIRTUAL RESELLING: EBAY

eBay is similar to Amazon in many ways. Amazon is recognized for retailing books and is diversifying into music, electronics, movies, etc. in both the new and resale market. Once the name recognition and systems for fulfillment are in place, it is a matter of Web page creation and product acquisition to expand into other markets. But whereas Amazon uses the industry standard for retail establishments (35%) as a pass-through markup for timely delivery of new books, eBay takes a different approach.

eBay is a matchmaker for buyer and seller, usually for used goods.[13] Responsibility for listing the goods, and for their quality and delivery, is the seller's. For providing this virtual meeting place and the bid arbitration between the seller and multiple buyers, eBay receives a small listing fee and a sales commission of about 5%. By listing over 1 billion items in 2004 for its 135 million registered users, the company reported gross revenues in excess of $3.2 billion. ***The company employs about 8,100 people averaging over 123,000 sales transactions per employee in 2004.*** Without technology, this number of transactions per employee would be impossible. But with technological magnification, this number becomes entirely possible and very profitable. In 2004, eBay's net income was nearly $400,000 per employee.

As eBay's infrastructure, methods, and customer base grow, the number of transactions per employee facilitated by technology will continue to increase. A possible techonomic metric for monitoring the techonomic performance of eBay is the transaction efficiency metric.

Transaction Efficiency TM

Assumptions:

1. Purpose is to measure organizational effectiveness in joining buyers and sellers online.
2. Progress in customer acquisition, productivity, and efficiency of operations should ultimately affect this metric.

$$TM_{\text{Transaction Efficiency}} = \text{TRANSACTION/EMPLOYEE}$$

where
TRANSACTIONS = Number of transactions arranged annually
EMPLOYEE = Number of full time employees at the company

Since eBay obtains products from so many different sellers, mostly in small quantity, it is not straightforward to negotiate favorable terms with each seller. eBay's approach was to create the PayPal network. PayPal allows eBay to emulate a credit card provider and profit from the cash float between the buyer's payment and the seller's reception of the goods. Floats are smaller and briefer than those enjoyed by, say, Wal-Mart — but the cumulative cash float becomes great as transactions multiply.

Not only does the eBay model sell products that are not in the physical inventory of the company, but it also outsources the virtual merchandizing of the product listing to the seller. By providing user-friendly listing tools to create product Web pages, eBay eliminates its own labor content in the listings, thereby increasing its margins as the quantity of listings increases. eBay receives a listing payment when the product goes up for auction and a sales commission when the sale transacts.

Bricks and mortar replaced by the virtual storefront. Ad layouts offloaded to the seller. Financial and fulfillment quality of the sellers monitored and reported by the buyers. Inventory owned and provided by the sellers. Product features determined by an educated buyer without paid sales staff. The world's largest flea market, eBay, is brought to you online by technology that is transforming old methods to create new and vibrant businesses.

The first three laws of techonomics combine to support the success of both Amazon and eBay:

1. The Law of Ubiquitous Computing provides low-cost, high-performance PCs for millions of homes and businesses. Buyers and sellers are equipped in increasing numbers.
2. The Law of the Ubiquitous Network interconnects these computers, creating a snowballing critical mass for buyers, sellers, and products. Buyers and sellers are connected in increasing numbers.
3. The Law of Increasing Productivity shrinks the size of organizations as transaction costs tend toward zero. Buyers have the tools to locate and

execute the sale without any other human intervention, supporting an amazing number of transactions per eBay employee.

Add in the economics of free cash flow by electronically controlling the transaction in the timeliest and most cost-effective manner for the company, and you have a new model for twenty-first-century business. The Franchise Effect (Chapter 5) is starting to grow around eBay as geographically distributed franchises build upon the eBay infrastructure to facilitate the posting of items on to the virtual auction site. Healthy organizations attract new partners into the expanding network of twenty-first-century techonomics.

VIRTUAL MEDIA: APPLE

First there were wax cylinders, then vinyl disks, then stereo long-play records, then eight-track tapes, then cassette tapes, then compact disks, then digital videodisks (DVDs), and now electronic files. *The techonomic process of layered innovation has been repeated in many fields: first there is joy in the discovery (the first phonograph), then distribution of the capability, then improvements in the quality (stereophonic), then the ability for the consumer to create the media (recorders), then miniaturization of the electronics and means for delivery (transistors), then virtualization of the delivery via a wired network and soon ubiquitous distribution via a wireless network.* The Apple iPod represents the latest in the line of discontinuous innovations within the music industry.

With each step, reproduction quality advanced, media durability improved, and replication cost diminished. With the advent of the MP3 music file format and the plethora of hardware players, the music industry is moving though a period of massive change. When technology first shifted customer habits from records to compact disks (CDs), there was little change in how the music industry did business. Artists recorded, publishers produced, industry marketers promoted music, and customers bought the physical end product (CD) at music stores or through the mail. The digital age began to change the industry when low-cost (First Techonomic Law) CD duplicators allowed customers to copy the music they wanted. The expanding network (Second Techonomic Law) allowed the music, legally or illegally, to be transmitted via files to anyone, anywhere. For enjoying music away from the computer desktop, few wanted to carry a portable computer or even a clunky and limited portable CD player. Enter the MP3 music player. MP3 stands for "MPEG-2 Layer 3," which is an audio compression standard developed by the Moving Picture Experts Group. This technology encodes digital audio in a space-efficient manner.

While most consumer electronics companies saw the obvious opportunity to develop a market for the MP3 players, Apple took a broader view and created an entire system for linking buyers to sellers, providing music mobility for the masses.[14] For the vast majority that had neither the technical expertise nor the immoral desire to pirate music, Apple created an integrated, easy-to-use system that gave buyers legal ownership of what they wanted: easily transportable music. The approach included software that allowed the user to digitize and load their music collection to the iPod (an MP3), download and pay for individual songs from a collection of thousands, and organize their own play lists as desired. The recording industry is forever changed.

Since inception, Apple has now sold, via download, more than half a billion song files. Each one of these files represents a transaction between a buyer (music consumer) and seller (music publisher or freelance artist) orchestrated by the invisible hand of Apple. Apple negotiates the sale price with the publisher, establishes the Web site parameters, and maintains the Web portal to the marketplace. Each $0.99 song download is a study in techonomics, with technology enabling free cash flow for a company that is transferring organized bits of information from the originator to the user. There is no cost of goods other than the initial set-up costs, minimal operating costs, and the royalty cost to the originator after the transaction is complete.

Even the iPod hardware unit itself is a study in short product life cycle and the value of partnerships. The iPod is a simple device, little more than a portable hard drive with input from a finger wheel and output to a small display and earphones. Not a lot of subsystems to advance. Since its introduction in November 2001, the iPod has morphed into the U2 iPod, the iPod Mini, the Shuffle, the Nano, and probably more by the time you read this. The 60-GB model lets you download up to 15,000 songs. Each product has variations of color casing, display, memory capacity, footprint, and battery life. What remains the same are the important file format, the interface to the computer, and the interface to the audio output. Vital as it is to upgrade performance as new generations of technology emerge, it is also important to retain consistency in the interface portals that tie a device to the world.

By maintaining consistent interfaces and establishing a strong leadership position in the market for MP3 players, the iPod has spawned an entire industry of complimentary products. Radio-frequency transmitters connect the iPod to any radio in a vehicle, home, or office, allowing users to become their own disk jockeys. Small speaker systems that have a dock for the iPod allow surround sound music to be taken anywhere as the centerpiece for any table. Necklaces have been designed to hold iPods. Digitizing microphones from third parties allow the iPod to become a dictating machine. Radio shows digitize their content for "podcasting."

Technology increases the product storage capacity and extends battery life, while allowing improved screen resolution and color displays for minimal cost increases. As a result, the iPod takes on new possibilities as a digital wallet for pictures and videos.

Where is the iPod headed? Will its technology be assimilated into cell phones like the Palm Pilot? Or will it make CDs a thing of the past in the same way that CDs made vinyl records disappear? What would techonomics predict? Techonomic trends indicate that the hardware price point will decrease due to Moore's Law and competition. Over time, the iPod's functions will be incorporated in other portable devices, most likely the cell phone, but as music is a form of entertainment, stand-alone units will also remain. Apple will be able to maintain a leading position if it continues to acquire and make easily accessible the media desired by the customer at a reasonable price.

As technology advances, anticipate an iPod that becomes popular as a digital movie player, an iPod that allows wireless Internet connection for downloads of music files, an iPod for wirelessly receiving radio podcasts, and iPods in automobiles.

Within 10 years or less, CDs will be a memory, because they are not the most cost effective way to distribute music. DVDs may see the same fate as distribution bandwidth increases, compression improves, and storage capacities continue to increase. These trends are already beginning today as record companies have seen their sales of CDs decline. Technology has permanently changed music distribution and enjoyment. It is on the cusp of doing the same for video distribution and enjoyment. Now the scramble is on for the economic model that will succeed in an industry that will perpetuate simply because people will always want to be entertained in the most convenient and inexpensive manner available.

EMERGING TECHONOMIC CONCLUSIONS

Time is money! Successful businesses use technology and strategic relationships to maximize free cash flow for their operations. Several factors contribute to structuring a business to maximize free cash flow. These include minimizing the time between obtaining an order and receiving payment from buyers, negotiating long-payment terms with suppliers for reimbursement of goods, minimizing or eliminating cost of goods, minimizing or eliminating inventory for supplies, minimizing or eliminating warehousing of finished goods, and using other people's capital for manufacturing and retailing facilities. Techonomic metrics like free cash flow per share or the number of transactions per employee allow the myriad of system advancements to be measured in the aggregate by a simple value. System adjustments that improve the metric advance the organization toward its operational goal.

Technology has changed the speed, complexity, and coordination with which commerce can be performed. Money now travels at the speed of light from buyer to seller, although due to negotiated relationships the money often does not travel as fast from seller to supplier. The buyer can now accomplish complex transactions, like the personal configuration of a personal computer. User-friendly, interactive Web pages eliminate the need for human salespeople to consummate every sale with a buyer. This significant sales labor savings is made possible by the ubiquitous availability of personal computers and the interconnected network of networks, the Internet. Next, the data handling capabilities of the network convert the buyer's desires into orders for the seller.

eBay has proven that the Web is capable of effectively matching hundreds of millions of buyers and sellers for the exchange of billions of items. Dell has shown that scores of suppliers can be coordinated to fulfill the desires of the buyer without human administrative intervention. Apple has created an automated information delivery system and infrastructure to match music producers with the desires of millions of music aficionados. Essentially, the iPod delivers the musical experience with convenience, selection, and economy superior to any previous system in history.

This section would be incomplete without mention of Google and its advancement gained from riding the twenty-first-century techonomic trends. By matching information seekers with the vast Internet reservoir of information, Google produces strong free cash flow by selling pass-thru links on its automatically generated virtual pages. ***Google epitomizes of the finder of "perfect information": free, instant, all encompassing.*** The only drawback is the overwhelming amount and the validity of some of the sources. Revenues are generated from sites that seek interested eyes for their materials. Prequalified viewers are passed to these sites based on the active inquiries that the viewers make to the Google search engines, a virtual, all-knowing finder for a small fee per transaction performing billions of information transactions a year (a month? a day?).

Today's fittest companies are also raising the value of annual revenue per employee. A decade ago, $100,000 in annual revenue per employee indicated healthy performance for a range of companies. Now, with outsourcing, vast supplier networks, and virtual relationships, the emerging techonomic companies are entering the range of 500,000 to $1,000,000 in annual revenue per employee. Even the best older companies, saddled with less agile corporate structures and remnants of the old economy, struggle to reach revenue per employee levels of half that value.

Twenty-first-century techonomic businesses treat the time value of information as one of their greatest opportunities for cash management. By shifting transaction costs and operational support to both the customer and the supplier, these companies manage their operating cash assets in such a way as to create positive cash flow as revenues increase. This allows lower margins by minimizing the cost of capital to fund operations. These companies also rely heavily on a network of outsourcing to provide mass customized customer service, just-in-time manufacturing, and on-demand distribution. Result: a larger-revenue-per-employee operational model than has typically been possible in the past. *Availability of computers in the consumer's home (Techonomic Law 1), combined with the interconnectivity of customers, businesses, suppliers, and distributors (Techonomic Law 2), has resulted in efficiencies (Techonomic Law 3) and purchasing options that are reshaping the rules of commerce.* Competitive pressures in the free market requiring the adoption of best practices for improved productivity have never been so great.

SUMMARY

EMERGING ORGANIZATIONAL PROCESSES

The free flow of information has created a "frictionless" economy, speeding transactions at a pace never before imagined or possible. Electronic currency and instantaneous access to information has created opportunities for new economic models (positive-growth cash flow, bits-as-the-product management), new production models (mass customization, just-in-time inventory management), and new operational paradigms (virtual storefront, instantaneous reference qualifications).

POSITIVE CASH-FLOW MANUFACTURING AND RETAILING

Combining traditional payment terms for suppliers (30 to 90 days after goods received) with electronic ordering and payments from customers (mass customization, electronic currency, inventory management), strategic manufacturers and retailers are able to create a positive cash flow while growing their operations. Minimizing inventory for work in progress and eliminating finished product inventory diminishes waste and improves competitive advantages (examples: Dell, Wal-Mart).

Predictable Antiquation

Unlike planned obsolescence of the past, where products were designed to fail after so many uses, predictable antiquation depends on the relentless advance of electronic performance predicted by Moore's Law to regularly recreate demand for ever-improving product performance (example: Intel).

Virtual Operations

In the techonomic era, a storefront is no longer brick and mortar, but a URL and a computer display. Money is no longer cash, but an electronic and instantaneous transaction. Products are not limited to the physical, but have a value added by information or data (examples: Amazon, eBay, iPod).

Selling Hardware as Portal to Media Sales

The industrial economic model of selling the razor at minimal cost in order to sell the razor blades at a handsome profit has been supplanted by the techonomic model of selling the player at minimal cost in order to sell the media at a handsome profit — only the media now requires virtually no cost of goods other than royalty payments because it is an electronic file (examples: video games, iPod).

Financial Management Includes the Time Management of the Transaction

In the age of techonomics, it is not sufficient to simply make a profit if your competition is also managing the transaction cash flow to their advantage. Electronic currency, order acceptance, and inventory management offer new ways to gain financial leverage from wise implementation of technology in operations. Successful organizations will study and implement these management techniques.

Key Terms

mass customization	positive cash flow
endeavor cash flow	earnings
free cash flow	just-in-time delivery
virtual retailing	rolling warehouse
planned obsolescence	predictable antiquation

QUESTIONS

1. The concept of positive cash flow from operations facilitated by technology is one of the key insights you should glean from *Techonomics*. Do the math on these two examples (questions 2 and 3) to understand the importance of the concept.

2. A company manufactures and sells computers through traditional retail channels. Retail price of computer is $1000, and distribution/commission to the store is 30%. Payment terms for store to manufacturer are 30 days after receipt of goods. Manufacturer pays its suppliers when their parts are received, and for this example, manufacturing is instantaneous (highly automated!). Total cost of goods for the computer is $400. Manufacturer turns inventory six times a year (inventory cost in 60 days prior to manufacture). Determine the manufacturer's cash flow needs on a per-unit basis (dollars times the days required). Determine manufacturer's profit on each unit without cost of capital. Determine company profit if capital cost is 1% per month.

3. A second company manufactures and sells computers over the Web, made to order, using just-in-time delivery supply chain management. The retail price of the computer is $1000, and there are no sales commissions other than cost of operating the electronic data network. Customers pay for their customized order when the order is shipped (same time as placing the order). Cost of goods is $400. Suppliers are paid 30 days after their just-in-time delivered parts are received. Manufacturer does not carry an inventory, but builds computers when they are ordered. Determine the positive cash flow generated by each order. Determine manufacturer's profit on each unit without capital cost. Determine the additional profit on each unit if cash flow provided by the transaction is invested at 1% a month.

4. A developer owning a key piece of commercial real estate may choose to build a speculative retail facility, hold the property, sell the property, or enter a long-term lease with a reputable retailer. Consider the following inputs to the developer's decision: current land appraisal = $1 million, current building costs = $200/square foot, unimproved land inflation rate = 8%, improved land inflation rate = 5%, rental annual increase averages 4%, possible facility size = 10,000 square feet. If the going rate for a 3-year lease is $20/sqft (annually), and a reputable company commits to a 20-year, fixed-price lease at $25/square foot, which of the four options would you pursue as the developer? Justify your answer in light of risks and reward for the developer.

5. Intel is possibly the first company ever to create a business model that deliberately "cannibalizes" itself every 2 years by advancing its own technology to the point of making its old products obsolete. This approach requires that the company remain a technology leader; otherwise, it does not lead the market curve. It must also maintain backward compatibility with its previous products; otherwise, the market volume needed to succeed is not available. Finally, it must assure that its cost of capital for new/better manufacturing facility can be substantially repaid within the short life cycle of the product. Intel has masterfully navigated these challenges over the past 30 years. Looking forward, the key question for this model is this: When does the increasing cost of the manufacturing facility exceed the economic payback of the microprocessor marketplace? Assume a $1 billion facilities cost to make a new chip and a $30 per chip

wholesale price, with a cost of goods of $3 (it is only sand!) and a 2-year shelf life before antiquation. How many chips need to be sold annually to cover the facilities cost? Are there other costs significant to determining economic viability?

6. The virtual storefront is expanding each year as the Internet expands and the buying habits of more users are influenced. Analyze and compare the major contributors to the cost of selling a product for a physical and a virtual retailer. A few starting thoughts to consider: physical retailer costs include facility, retail labor, inventory, utilities, advertising; virtual retailer costs include Web site, inventory, shipping labor, shipping costs, advertising. Which approach benefits most from increased quantities of scale? From techonomic trends?

REFERENCES

1. Collins, Jim, Good to Great: Why Some Companies Make the Leap ... and Others Don't, Collins, 2001.
2. Dell, About Dell, 2006, http://www1.us.dell.com/content/topics/global.aspx/corp/en/home?c=us&cs=&l=en&s=corp.
3. Wikipedia, Pareto principle, Wikipedia, The Fee Encyclopedia, 2005, http://en.wikipedia.org/wiki/Pareto_principle.
4. Wal-Mart, Our Company, Wal-Mart, http://walmartstores.com/GlobalWMStoresWeb/navigate.do?catg=1.
5. Fishman, C., The man who said no to Wal-Mart, Fast Company, 2006, Jan/Feb, 67.
6. U.S. Census Bureau, U.S. Trade in Goods (Imports, Exports and Balance) by Country, Foreign Trade Statistics, U.S. Census Bureau, 2005, http://www.census.gov/foreign-trade/balance/.
7. Susette Kelo et al. v. City of New London et al., State of Connecticut Judicial Branch, 2004, http://www.jud.state.ct.us/external/supapp/Cases/AROcr/CR268/268cr152.pdf.
8. Walgreen Co., Our company, Walgreens.Com, http://www.walgreens.com/about/default.jsp?headerSel=yes.
9. Intel Corporation, About Intel, Intel, http://www.intel.com/intel/index.htm?iid=HPAGE+low_about_aboutintel&.
10. Microsoft Corporation, About Microsoft, Microsoft, http://www.microsoft.com/mscorp/info/.
11. Amazon.com, Amzn investor relations, Amazon.com, 2006, http://phx.corporate-ir.net/phoenix.zhtml?c=97664&p=irol-IRHome.
12. Amazon.com, Annual Report, 2004, http://library.corporate-ir.net/library/97/976/97664/items/144853/2004_Annual_report.pdf.
13. eBay, About eBay, eBay, http://pages.ebay.com/aboutebay.html.
14. Apple Computer, iPod, Apple Computer, http://www.apple.com/ipod/.

7 Emerging Techonomic Trends

Trends are bottom-up, fads are top-down.

— John Naisbitt, *Megatrends*

INTRODUCTION

The first three laws of techonomics provide a foundation for analyzing other developments currently shaping our society. When two or more of the first laws combine to support an emerging endeavor, that endeavor will likely become economically viable and find widespread adoption in the future. We are now seeing several endeavors being shaped into techonomic trends because they are highly favored by these laws. These techonomic developments are active in all four sides of the organizational square: energy, computation, communications, and community. This chapter discusses the marquee trends anticipated in the next two decades categorized by the organizational square and specific technology developments to watch as trend leaders (see Chapter 9 for techonomic trends in the more distant future).

ENERGY: JOURNEY TO RENEWABLE ENERGY RESOURCES

To understand contemporary techonomic effects on the direction of society, one must first understand their effects on energy. It was steam power that ushered in the Industrial Age, replacing the animate labor of the Agricultural Age with machine power. The Industrial Age rapidly developed based on fossil fuels — a nonrenewable source — magnifying our muscle. Society flourished materially, but the resulting pollution from combustion of finite fossil fuel reserves raises serious energy production questions in the postindustrial era. Answers to energy questions must be based on a holistic picture: technology, economics, politics, the environment, and society.

The key question to answer is this: What is acceptable risk for provision of energy? The way each nation/society answers this question over the next decade will set the course for their economic viability over the next 50 years. This is because it takes a long time to design and construct major power generating facilities, and worldwide competition for energy resources is increasing. Intelligent national leaders recognize the need to plan and develop energy generation capacity before crisis arrives rather than waiting for catastrophic shortages that demand emergency measures.

Societies need energy to function, and twenty-first-century societies worldwide need energy in increasing quantities to support improving standards of living. While conservation efforts are laudable, society can no more save its way out of an energy need than an individual can starve his/her way out of a nutritional need. Individuals function on food, societies on energy. Remove energy, and the standard of living is instantly diminished (remember what happens when we are without electricity for an hour, day, week, longer). This section considers various avenues of energy production readily at hand or eminent, not futuristic possibilities with no time horizon for their broad commercial application.

For this broad discussion, renewable energy sources are categorized into four broad classifications: biological (ethanol from plants, etc.), cyclical (solar, wind, wave, geothermal, etc.), chemical (batteries, fuel cells, hydrogen cycle, etc.), and nuclear (breeder fission, fusion, etc.) Techonomics balances the technology trends with the economic realities of the marketplace to give insight into the near-term future of this industry.

BIOLOGICAL SOURCES

Biological energy sources have always been available; wood and other organic products burn, releasing energy. It has been a grow-your-own energy plan! Now we can produce substances like ethanol, a substitute for gasoline that is produced from biomass originating as corn, sugar cane, seaweed, etc. As with any energy source, production and use of ethanol comes with both positive and negative considerations. Arguments favoring ethanol include production from locally grown crops (supports the economy), cleaner burning than gasoline, and little requirement for changes to the current fuel distribution infrastructure. Negative considerations include the fact that biomass combustion releases carbon dioxide (greenhouse gas) and that ethanol manufacture requires more energy than the product itself later releases.

This latter problem is documented in an article in the *San Francisco Chronicle:* "Production of ethanol from corn consumes more energy than it produces."[1] Additionally, some of that energy consumed is energy from the very fossil fuels that ethanol is seeking to replace. No amount of techonomic progress is going to change portions of the conversion process that require heat to decompose biomass into its component parts. Simple techonomic analysis exposes the ethanol cycle as little more than a government subsidy, "subsidized food burning," in the words of Cornell University professor David Pimentel.[2] The economics of requiring more energy to convert the raw materials to a useful form than is produced in the effort results in a system that would not be financially sustainable in a nationwide implementation. In the realm of politics, such an approach may strengthen ties to an important special interest group, but it is not a techonomically feasible alternative.

CYCLICAL SOURCES

Cyclical energy sources (wind, wave, solar, etc.) are a common consideration for renewable energy. In the broadest sense, the sun is the ultimate source of all energy on the planet (energy "under" the planet is another category, as we shall discuss

Emerging Techonomic Trends

momentarily). These cyclical energy sources do not produce carbon dioxide, hence protecting the environment. These sources are readily available in some form at most locations on the planet. Their conversion into electricity is also technically feasible. But currently, the economic viability and generation capacity of these sources limits them to a minor role. Economic challenges arise in the following areas: high initial capital costs, high operating maintenance costs, backup generation coverage (sun does not always shine, wind does not always blow), and large land requirements (because of low conversion efficiencies).

Wind Sources

The economics of wind power deployment can be explored in a detailed spreadsheet available from the Windustry project.[3] This analysis shows that operational costs exceed the value of the electricity generated at current prevailing prices, no matter how small the initial capital costs are. This is an example where government subsidies can make an industry a reality, but in the long run that industry will be only a small portion of the energy supply — and an economic drain on other resources.

Wave Sources

Wave power is less developed and even more of an economic uncertainty, suffering from reliability challenges related to harsh chemical environments (salt water) and low energy conversion efficiency. Without major technology breakthroughs in concept and execution, techonomic trends do not play in favor of these energy sources for the foreseeable future. Geothermal sources are in a similar state of early development. A few demonstration projects have realized the potential of generating power from heat trapped within the Earth in the form of molten materials. The technical problems with materials and the inability to determine the total amount of energy available at a given depth and location limit the economic viability of this source.

Solar Voltaic Sources

The phenomenon of electricity generated from photo-voltaic cells was first discovered by Frenchman Edmond Becquerel in 1839. In the 1950s, Bell Laboratory developed and improved photovoltaic cells for space.[4] Solar electric energy production from photovoltaic cells has been applied in remote locations and experimental installations, but this approach has not yet gained wide acceptance. Advantages are electricity produced without carbon dioxide production and no operating costs beyond initial capital to install. Drawbacks include large initial capital cost relative to the power produced, intermittent capacity based on available sunlight, and large collector space needed for useful power output. According to the company Nanosolar, *the current economics of solar electric generation technology are about three times the cost of electricity generated from fossil plants*.[5]

The foundational laws of techonomics favor solar cells looking to future trends. A breakthrough is needed in production cost of the silicon per unit area to reduce cost of solar cells to a competitive level. Traditional Moore's Law progress has focused on putting more components on the same surface area of silicon. Solar voltaic economics seeks to reduce the cost per unit area for the collecting silicon — a different problem that has not benefited from advances via Moore's Law.

The Nanosolar Web site provides an excellent techonomic analysis of the key metrics and the breakthrough approach they are pursuing to address the issues. Combining manufacturing approaches from the publishing and semiconductor industries, this company seeks to reduce the basic material costs by an order of magnitude, making solar voltaic electricity generation economically competitive in the foreseeable future. This analysis also reveals a thorough understanding of two performance keys that must be optimized for success: cost per unit area and generating capacity per unit area. The goal for economic viability is $0.50 per watt for a sheet of this material. There is no need for subsidies or tax credits at that level of performance because the economic productivity is competitive with market alternatives.

Techonomics anticipates advances that will meet the challenge posed by intermittent power generation associated with sunlight. The cost of electronics to synchronize solar-generated electricity with the utility power grid will significantly reduce as these distributed systems become popular. The development of more effective energy storage devices will also support solar electricity load leveling (ultra-capacitors, new battery composition). While an area the size of Vermont would be needed to supply America's homes with solar energy, the area is already available in a distributed fashion atop the homes where the power is needed. Techonomics anticipates a breakthrough of significant proportions as electrical generation transitions from centralized to distributed, much like computing did with the invention of the personal computer. If the economics of solar cells becomes competitive with central production facilities, there will be a major shift in the sources of energy supply for personal use within the next two decades.

CHEMICAL SOURCES

Chemical energy sources that produce or store energy without using the carbon cycle (production of carbon dioxide) include fuel cells, batteries, and the hydrogen cycle. Batteries and the hydrogen cycle are simply storage and retrieval means for energy. They can support the important mobile applications (vehicles), but due to conversion inefficiencies, they consume more energy than they produce. From a root techonomic analysis, they become supporters of the main generation methods in the overall system, as in a hybrid vehicle, but cannot serve as the fundamental generation method. Fuel cells are similar, but they can be configured with so many different chemicals and catalysts that it is difficult to review them in the broad view without studying a specific system. Suffice it to say that the reacting chemicals have to be separated and refined via energy input to create a fuel cell that has portable power capacity.

Batteries, or a similar energy storage system based on ultra-capacitors or fuel cells, will play a key roll in the rise of hybrid vehicles for transportation. Optimal operation of the internal combustion portion of the vehicle, combined with regenerative braking, allows hybrid vehicles to improve gas mileage from 25 to 50%. Such improvements economically pay back the differential equipment costs within the useful life of the vehicle at current fuel prices. Batteries will also be needed to supplement solar cell systems as a means to balance the electrical load between sunlit and dark periods.

The world has waited for major battery breakthroughs since the days of Thomas Edison, but only minimal progress has been made. There are completely radical approaches to batteries that may improve capacity, reduce charging time, and extend useful life called "ultra-capacitors." These ultra-capacitors store energy as charges in nanometer-sized features within porous materials. Since they do not use chemical reactions associated with batteries, they do not degrade like a battery does. They can be rapidly charged, as they do not require a chemical diffusion process to promote charging. They can provide tens of thousands of charge/discharge cycles (hence improving the economics of life-cycle costs). In an ancillary way, ultra-capacitors benefit from Moore's Law as the demand for better batteries for mobile devices has resulted in accelerated research in this breakthrough technology.

NUCLEAR SOURCES

The final possibility on the horizon is nuclear energy. Fission and breeding fission are the nuclear-power-generating possibilities for the foreseeable future. Fusion has been 50 years away from commercial viability for the last 30 years. It remains 50 years away. Fusion suffers from a host of material, control, and thermal issues that must be solved exclusively for its success and therefore cannot greatly benefit from the march of technology in other fields. A techonomic appraisal offers little confidence that fusion will be a viable energy alternative in my lifetime or that of my young children.

Contemporary energy trends are obvious, even if some nations seek to deny them. As previously noted, nuclear fission generates more electricity worldwide than was produced by all energy-generating methods in 1961. In the 50 years since the first commercial, nuclear-powered electric generating plant came on line, nuclear energy has grown to provide 16% of the world's electric power. As the techonomic review of history revealed, such rapid growth is the result of a techonomic trend favoring a new, more powerful and productive means (nuclear) over past methods (oil, coal). If the nuclear industry had experienced a consistent and realistic regulatory environment, it could now be producing over one third of the world's electricity without producing greenhouse emissions. With fuel reprocessing methods, the useful supply of nuclear fuel can be extended. And with breeder reactors, it can be renewed indefinitely.

Using techonomics to anticipate future trends would indicate that nuclear energy will proliferate. In the long run, the economics of an application will overcome government regulations and irrational, special-interest arguments — it is a matter of economic pressure. In the years ahead, nuclear power will "win." Why? Consider these assumptions:

1. People have consumed energy in increasing amounts throughout history.
2. Standard of living is linked to energy consumption.

3. World energy consumption will increase rapidly in developing nations. Even if the developed countries reduce their growth of energy usage, total world requirements will continue to grow rapidly as a larger percentage of the world's population seeks better living standards.
4. Fossil fuel reserves are limited, and they inevitably produce greenhouse gas emissions. Ultimately, other fuel sources will replace them, but renewable sources (wind, solar, biomass, etc.) do not provide the energy density required to meet the world's power demands currently or in the imaginable future.
5. Deployment obstacles to nuclear power can all be overcome. They are typical of the early commercial adoption of any new technology, but the ramifications are larger because the energy density (the concentration of energy) is orders of magnitude greater than other fuels.

Political obstacles require public support to overcome. As France has demonstrated, a major shift can occur in less than 20 years if political will dictates. This shift will occur in other places over the next 25 years as the combined economic uncertainties of petroleum supply, global greenhouse emission regulation, and particulate/chemical pollution force the use of more reliable and cleaner alternatives.

Despite this inevitable development, I would by no means encourage a U.S. utility to invest in nuclear capacity during the current politicized regulatory climate. The uncertainties of shifting regulation and incredible liability insurance premiums have to be significantly reduced for a rational investment to be made. Remove these economically debilitating moving targets, and the techonomic metric showing relentless growth in global per capita energy indicates nuclear electric power generation in the U.S. will again advance as it already is in the rest of the world. Until that happens, the U.S. will see ever-increasing energy trade deficits as it competes, in an expanding worldwide energy market, for the steadily diminishing reserves of petroleum, natural gas, and low-sulfur coal.

CONTEMPORARY ENERGY TECHNOLOGIES

Three contemporary energy technologies are benefiting from one or more of the fundamental trends of twenty-first-century techonomics. These technologies and how they are aided by the trends include:

- **Electricity generation by nuclear power** — Techonomics predicts that nuclear fuel will be the fuel of choice in decades ahead if free market forces prevail. Digital control systems (First Law) and sensory networks (Second Law) continue to advance far beyond the capabilities of generation plants designed over 30 years ago. As we use nonrenewable fossil fuels in increasing quantities and their supplies diminish, the economics of fuel costs will shift the operation advantage toward nuclear power. The environmental concern of fossil combustion vs. nuclear fuel storage begs the question of which issue is more manageable. Is it easier to contain today's carbon dioxide gaseous emissions or to transport and store the

spent solid fuel materials for generations? Both are challenging tasks, but as society experiences the observable effects of rising fuel costs and pollution from combustion, the nuclear option will become more viable.
- **Breakthrough in solar-voltaic electricity.** Continuing reduced cost of digital electronics (First Law) will provide an inexpensive means to synchronize solar-voltaic electric generation with the power grid. Fabrication methods leveraged from electronics, coupled with new materials and production approaches, will reduce the cost of solar-voltaic panels. It is not so much that solar panel efficiency must increase; cost of panel per unit area must be reduced to reach economic viability. *The power network will begin transforming like the computational network from central, concentrated production to distributed, localized generation (Second and Third Laws).* Solar power will augment centralized base-load generation, driven initially by pollution credits and ultimately by the economics of the technology in mass application.
- **Proliferation of hybrid vehicles.** Hybrid vehicles that combine electric and internal combustion power sources are emerging in the marketplace. *Techonomic fundamentals predict that this time, hybrid vehicles will succeed in the marketplace.* In contrast to earlier efforts of total electric vehicles with limited range performance, the emerging hybrids have good performance and efficiency to match. Computer controls (First Law) have advanced, and their cost has been reduced. The 3000 to $4000 additional hybrid vehicle price is recouped in a 4-year period via reduced fuel costs, particularly if the usage patterns are typical of city driving (stop and go).

Mass production of hybrids will reduce this cost differential, and rising fuel prices will further diminish the economic payback period required for the customer to justify the investment. Note that hybrid vehicles have a near-term market penetration advantage over hydrogen power, because the fuel delivery infrastructure for their operation is already in place. This certainly is not the case with hydrogen. Whenever infrastructure has to be modified, particularly across an entire country, it is difficult for new approaches to be embraced, because no one wants the economic risk of the new infrastructure (to deliver hydrogen) while they wait for the demand to grow (the number of vehicles using the new fuel).

Economically speaking, the hydrogen has to be produced from other energy forms, most commonly the separation of water into hydrogen and oxygen by electricity. This energy must come from somewhere, at a cost. If the energy for this originates from electricity generated by fossil fuels, the result is more air pollution from the total cycle and more fuel cost due to inefficiencies in the total cycle. Hybrids are a strong incremental step while a long-sought breakthrough in alternative fuels is pursued.

Summarizing, we have looked at biological, cyclical, chemical, and nuclear power generation sources for our growing electrical energy needs. A generalized techonomic review reveals that biomass has limited potential due to its negative productive energy balance. Of the cyclical sources, solar voltaic cells offer the greatest potential if unit area material costs can be reduced while maintaining electric conversion efficiencies. Wind, wave, and geothermal generation systems neither benefit from the fundamental techonomic trends nor are they currently economically viable. They are destined to experimental deployment, meeting a small percentage of power needs, with new deployments being fought by area residents to prevent eyesores on the land. Chemical generation methods, other than fossil, primarily serve as storage and retrieval means for electricity. As such, they will not be the fundamental generator, but the load leveler. The emergence of the ultra-capacitor may resolve economic issues for larger power storage for load leveling (vehicular and residential power). Finally, nuclear generation shows technical promise constrained by political uncertainty. Cultural acceptance of reasonable risk for all energy forms, as opposed to the current low tolerance for certain energy forms, is needed to spark the nuclear power industry's rebirth. As nonrenewable, high-polluting, carbon-based energy sources are further depleted and nuclear technologies advance, anticipate a continued worldwide trend toward nuclear electricity generation.

COMPUTATION: ALL THINGS DIGITAL

The advent of electronic computation and mass information communications is causing another age to dawn. Calling this age the Information Age would be like calling the Industrial Age the Steam Age; information is only one key part of the picture. For lack of better terminology, this age has been referred to as the postindustrial or postmodern era. Perhaps a better term is ***the Virtual Age: an age in which digital computation, the ubiquitous network, and increasing personal productivity allows virtual representation of all that was once physically tangible.***

As we view organizations through the simple four-square analysis — energy, computation, communications, and community — it is evident that advances in computation and communications dominate the current transition beyond the industrialized age. The spotlight in this age is on the computer. Society's current journey is "from atoms to bits."[6] It is an economic transformation. The old economy measured the cost of goods in terms of intrinsic weight, while the new economy measures the cost of goods based on the intrinsic value of the function they perform.

New goods and services give rise to new occupations, new businesses, and entirely new ways of organizing society. Pictures, documents, money, video, music, telephone calls, entertainment, and records of all types have migrated from analog or atom-based media to digital representations. In the digital form, any data type can be stored and computed upon with the same electronic hardware. There is no need for multiple information and media formats for video storage, audio information, photographic information, written information, etc. The universal world of data storage, management, and manipulation has arrived. The information is digital, and its storage is electronic memory.

In digital form, information can be shipped anywhere at the speed of light. It can be transformed, manipulated, searched, stored, retrieved, shared, or massively distributed, all within the virtual world of the computer and the Internet. Today, actions on information can be performed instantaneously, simultaneously, anonymously, ubiquitously, and inexpensively. Labor transformation will be as significant in this age as it was in the move from Agricultural Age to Industrial Age. Steam power supplanted (to a large degree) physical effort in the previous transition. Computational power will "supplant" mental labor in the current transition. *The long-held promise of automation has been realized in this generation in the form of a personal computer capable of replacing bureaucratic labor at a fraction of the cost.*[7]

Franchised business models replace thinking positions with simplified repetitive processes that can be quickly learned and accurately repeated. Tougher assignments like inventory management and financial accounting operations are automated by the network of computers in everything from the cash register point of sale to the delivery truck global tracking system. Those mental labor functions that the computer cannot fully automate can be quantified and transported worldwide, to be performed by the most cost-effective labor pool.

The move from atoms to bits means complex, effective models of reality are now resident in computers, and the line between real and virtual is blurring. Money is electronic, consummating financial transactions at the speed of light — no more mail float at the end of the month. Fluency and ease with the computer is the technologic skill required for this economy. The labor of thinking for the working masses — that is, of "task-oriented" mathematical thinking and any systematic, repetitive thinking — is diminished. Computers, and their increasingly sophisticated software, do this kind of thinking for the user.

Computers are superb at doing the math (spreadsheets, making changes, etc.). They write reports from automatic measurements (financial reports, applications, etc.). They search the available media for your preferences (search engines, Tivo, etc.). The information worker need only follow the script. Work ethic, computer literacy, and compensation requirements are the economic measures of workforce competitiveness. Human labor in this specific context is not particularly creative or interesting, but the efficiency of this system will free human time for other pursuits.

Here are a few of the computational product trends gathering momentum from twenty-first-century techonomics:

- *Digital omniscience.* The big brother to data is knowledge. All electronic transaction data is available on the network of networks (Second Law), and the computational power to analyze it is rapidly progressing (First Law) without the need for human intervention (Third Law). With the volumes of data being collected from transactions on the ubiquitous network, the companion advance will be intelligent systems that mine useful information from this data. Consider how much the network knows about you. Every time you use your credit card: electronic transaction; every time you write a check: digitized transaction; every time you write an e-mail: electronic record; every time you search the Web: cookies watch; every time you make a telephone call: electronic transmission. Following your electronic data footprint reveals a lot about you. Cash transactions are not the answer, either. Radio Frequency Identification (RFID) devices are becoming so inexpensive (First Law) that they will be embedded into major currencies within the decade. Amazon.com has been an early mover in trying to understand customer preferences and use that understanding to increase sales to that customer. Such buying recommendations are the tip of the iceberg of data mining for specific value added. In the fight against terrorism, automated monitoring capabilities and data mining for all forms of personal communications have been enhanced and used. We face a brave new world in which not only every transaction is captured, but the computational capability exists to infer potential future behavior from combinations of those transactions. Certainly, the data and analysis means are available if the models of human behavior become more accurate.

- *Genesis of the virtual world.* This generation will witness the creation of a virtual world for interaction, entertainment, commerce, and education. First Law computation advances make graphic displays realistic, compelling, miniaturized, low power, and inexpensive. The Second Law joins the world, via the Internet, into a virtual community only a keyboard away, or in a totally wireless world, provides the virtual anywhere. Three-dimensional animated video games create worlds so compelling that players spend all their waking hours in an imaginary environment. Animated movies blur the line between the real and the computer generated.

Each year, technological advances in hardware, software, and content further blur the line between real and virtual. Today, a greater diversity of physical goods can be accessed via the Internet virtually than can be made available physically at the largest real shopping mall. The eminent missing product, which will link the masses with a compelling virtual world, is the transformation of the computer display to a head-tracking, eyeglass-like immersive device. Such a device could lead to the creation of totally virtual life experiences rivaling the experience of the real world. A comfortable system combining high-resolution, wide-angle stereo display, seamless kinesthetic tracking, and directional stereo audio output would immerse the user into 90% of the real-world data input of the human sensory experience. In the years ahead, such devices will change our comprehension of the media experience. With techonomic trends supporting electronic miniaturization, community network interaction, realistic generation of animated displays, and a market demanding an ever-more-realistic entertainment experience, the age of the virtual world is approaching.

COMMUNICATIONS: EXPANDING CONTROL AND INFLUENCE

Combined with the powerful leverage of the computer, the growing network of digital communications provides an expanding breadth of control in the Virtual Age. Management, via electronic networks, can be *virtually* anywhere. Fading away are organizational charts with layers of middle-level management to carry out executive directives. While vestiges of these organizations still exist, primarily in older companies with their roots in the industrial age, there is a measurable increase in the span of control in Virtual Age companies. It is evident from the techonomic metric, described earlier, of "revenue per employee." It is also evident in the vernacular of the last decade: right sizing, lean engineering, flat organization, remote operations, etc.

Free market economic pressure is the companion of technology advance causing this sweeping organizational change. Companies must not only produce good products, they must use the most efficient methods of production, distribution, capital management, and customer service to provide their products/services at a competitive price. The organization and use of labor resources become increasingly critical to the continued economic success of an enterprise. Information technology has become the tool to optimize labor use, just as mechanization was the tool that transformed labor patterns from the agricultural to the industrial age. Organizational structures must change, or the organizations themselves will become extinct.

Ubiquitous communications in the form of high-bandwidth digital connectivity is opening the possibilities of expanding remote operations. Plant engineers for Tennessee Eastman operate plants across the world, remotely, by observing plant performance via distant sensors and making adjustments to processes a world away. The University of Phoenix and the National Technological University train thousands of graduate students across the nation by satellite and Internet links rather than in brick-and-mortar classrooms. You gain access to more information than is available in the Library of Congress many times a day simply by making a query on Google or other search engines on the Web. Not only has there never been a period in the world's history where so much information was being generated, but there also has never been a period when that information was accessible in a nonlinear, "searchable" manner by the masses. The resulting shifts in organizational structure are only beginning to emerge and will continue to develop as the Virtual Age matures.

The foundational laws of techonomics — continued electronic intelligence, continued network expansion, and continued organizational efficiency — anticipate several organizational trends including:

1. **Organization structure will flatten.** A broader span of control, supported by communications and information structures, will grow simultaneously with reducing layers of management and bureaucracy in all organizations driven by the private sector. The pressures of competition will force nonoptimal (price and productivity) labor use to be eliminated.
2. **Virtual companies will proliferate.** Small leadership teams will commandeer the needed resources, wherever the best are located, and manage them remotely using electronic commerce methods. The term "micromultinationals" has recently been coined to describe technology startups that obtain capital with a small domestic team while most of the development labor is scattered around the world. One key to success is to retain the "crown jewels" of the organization as outsourcing occurs, or the day comes when the organization has no sustainable competitive advantage.
3. **Franchise structures will continue to grow.** Within this growth structure, a successful organizational model is conceived, tested, and proven on a local scale and then rapidly distributed worldwide with electronic checks and balances to monitor progress and foster success.

4. **Global labor rates will equalize.** Availability of inexpensive global communications breached the last natural barrier to large differential labor rates for information workers. Twenty years ago, a transpacific call could easily be $1 a minute ($60/hour). Today, it is under $0.05 a minute ($3.00/hour). The international labor rate communications differential for information worker jobs is now very small (thanks to technology advance, communications deregulations, and capital investment in building the networks). Indirect labor costs now become important in determining any competitive labor rates. Healthcare, retirement, unemployment insurance, workman's compensation insurance, and liability insurance become significant labor considerations as the communications costs diminish to insignificance. Hence, the U.S. consumer is serviced by telemarketing calls from India, architectural drawings from Taiwan, software programming from Ireland, manufacturing from China and Japan, and accounting assistance from Singapore. Traditional barriers to entry for skilled labor have been eclipsed by the ubiquitous network and virtual digital workflow in the Virtual Age.

5. **World languages will converge.** The Internet, television, and other technologies are healing the curse of the Tower of Babel. Today, humankind's ability to work together is being restored on a large scale, and the return of "universal language(s) is the key. Because of TV, verbal language accents in the U.S. have become more homogenized. Computer programming in English language (COBOL, FORTRAN, BASIC, C++, HTML, etc.) has caused the technologically elite in many countries to learn English. Commerce and communications today are crossing more international boundaries. According to Manfred Sellner, assistant professor of linguistics at the University of Salzburg, "English, formerly perceived as a symbol of linguistic imperialism, is now accepted as the primary vehicle of economic globalization."[8] Techonomics anticipates a trend toward unification of languages as more people become bilingual in one of the languages of major population groups for commerce: Chinese, Spanish, and especially English.

Whether the job is software development, telemarketing, product service call center, accounting, drafting, or Web site design, the digital pipeline moving at the speed of light is not concerned about the source or destination of the endeavor. *In the Virtual Age labor cost, cultural work ethic and trainability are more important than societal infrastructure (roads, water, utilities, etc.) as long as the digital pipeline is available.* In some regions, the first installation of the digital pipeline will be wireless, reducing the capital and land "right-of-way" requirements for the infrastructure while leapfrogging the wired infrastructure of other nations. Hence, another past U.S. labor protection barrier, the great physical infrastructure for support of commerce in the U.S., is leapfrogged by these innovative communication technologies.

This competitive pressure is accelerating industrial evolution into the Virtual Age at an unprecedented rate. The economic barriers to significant labor rate differential for mental labor have been removed by the onslaught of technology in the digital age. A few of the key product trends in communications to note include:

- **Wireless world**. The combination of diminishing electronic costs (First Law) and increasing network connectivity (Second Law), along with the desire to be more efficient (Third Law), is leading to the rapid expansion of wireless networks. Cell phones have blazed the path, and mobile computers are rapidly following. RFID tags will enable every package/product of any significant value to be tracked from cradle to grave, and any useful information (maintenance, owner, location of manufacture, location of use) will travel with the product. It will be cheaper to automate this process, having the information always available, than it will be to maintain any written records of the product history. *Anticipate an expansion in the Internet protocol numbering system so every manufactured product will have its own address.* The RFID communicates to a unique Web site for every significant product you own, tracking usage patterns, location, maintenance records, etc. Expect RFID in passports, too, following people across every border. Blue Tooth technology will allow handheld devices to communicate with each other while they also communicate with any intelligent device requiring transactional information (vending machines, toll plazas, checkout registers). The wireless world will complete the promise of 7/24/365: anyone, anywhere, anytime. The access to products and information that companies like Amazon and Google have brought to the desktop will be available ubiquitously. Instant gratification just keeps getting more gratifying, instantly, in the Virtual Age. The real-world delivery companies (United Parcel Service, Federal Express, etc.) should see continued expansion as they become the physical fulfillment channel for virtual commerce.
- **Media convergence: home, mobile, vehicular.** Convergence is happening in three key areas today: home media, hand-held devices, and vehicular media. The pockets of our generation are filled with small, electronic gadgets, thanks in no small part to Moore's Law. Cell phones, digital cameras, personal digital assistants, video cameras, MP3 players, personal dictation devices, web browsers, e-mail sending and receiving devices, Global Positioning System (GPS), and Blue Tooth activating keys are cluttering our pockets and countertops. The "Swiss Army Knife" of mobile convergence will emerge as the device that joins many of these functions in a small, reliable, high-performance package. Many cell phones now take pictures, include e-mail, track GPS, play music, and so on. The next decade will bring an army of product offerings that seek to find the market optimum between functionality, price- and ease of use. *As chip performance increases (First Law) and the wireless network expands in coverage and capacity (Second Law), the ability to package more media forms into the same package will continue.* The question remains: What is the market interested in, and what price will the market bear? The successful business models will arise from the hardware/service models

that currently dominate the cell phone market. Such an approach allows the high-entry-level cost of new, cutting-edge technologies to be distributed over several payments of a monthly service agreement. The resulting residual service payments create predictable revenues, typically with strong profits, once the capital of the infrastructure is recovered.

- **Vehicular media expansion.** With the advent of GPS in automobiles and trucks, color displays in vehicles are now becoming commonplace. The conflict between useful service and driving distraction is growing. Convergence in the automobile will bring together the same devices as the handheld convergence, with the addition of a screen large enough to provide maps and entertainment. These systems are now available in high-end vehicles and will progress into mass applications as price falls and the features and network increase. Anticipate widespread use of mobile text messaging, web browsing, MP3 interfaces, DVDs (on rear screens), integrated GPS, integrated mobile phone coverage, and a smart sensory system reporting vehicular conditions to "big brother" in the event of an emergency. As the network of intelligent sensors on our major roadways is expanded, anticipate communications of that information and of weather information directly to the vehicle. Honda has developed an interactive path-planning agent that takes traffic information into consideration. Also anticipate a host of safety sensors that monitor speed and following distance, alert when highway sideline is crossed, and monitor road conditions to recommend safe speeds. The network of networks, combined with wireless access, will provide traffic information, weather information, hotel availability, make dinner reservations, and read your e-mail to you as you drive. Most of this is already available in vehicles as optional products, but it will soon follow First and Second Law trends to the mass market.

- **Ubiquitous sensory network.** According to *One Digital Day,* you should carefully consider how you look in New York City, because you are captured on video over 20 times in the typical day.[9] Video cameras in elevators, lobbies, restaurants, convenient stores, gas pumps, street lights, etc. are capturing and saving, at least for a while, your image. ***Camera and storage costs have plummeted due to Moore's Law, and remote cameras can send information to a central repository via the expanding network.*** Cameras do not just track us; they track our transactions with the aid of the network. Federal Express captures a digital image of every package it ships (over 3 million a day in the Memphis processing center) and retains this information for months in case of a delivery question. Major highways have sensors to monitor and redirect traffic flows and control traffic lights. New vehicles have global positioning sensors that track location, speed, and perhaps even cargo manifests. Cellular telephone towers will be equipped with sensors to monitor the air quality to provide early warning for pollution alerts or terrorism acts. All the foundational laws of twenty-first-century techonomics point to an

ever-expanding network of sensors, data collection, data storage, and automatic evaluation of our every act.
- **Automated bureaucracy.** Although this may sound like an oxymoron, the automated bureaucracy is growing at an increasing pace. A combination of high labor costs (economics) and advancing computer capabilities (technology) has made the replacement of human mental labor economically possible and a competitive necessity. Touch-screen systems are eliminating the need for bank tellers, post office clerks, cash register attendants, and tollbooth operators. Automatic telephone systems eliminate the need for receptionists and greatly reduce the number of personnel needed to field product inquiries and technical support. Automated form processing for everything from loans to computer orders streamlines the labor content. Look for more evidence of the automated bureaucracy in retail stores as they trim costs to compete with the most efficient purveyor of automated bureaucracy, the Internet.
- **Personalized pharmaceuticals.** Many advances in the understanding of DNA and metabolism have been made in the last decade due to new imaging instruments (First Law), massive data processing (First Law), and collaboration (Second Law) in the research community. The field of personalized medicine, particularly personalized pharmacology, is emerging. This refers to the development of pharmaceuticals and treatment methods based on the DNA of the individual being aided. Rather than a blanket approach to a given disease, these treatments take into account the DNA configuration of the patient to determine the treatment option most likely to succeed, even to the point of creating drug delivery "tags" based on DNA from the patient. Without the massive computational effort to identify the human genome in the 1990s, the information needed to develop these treatments would not be available. The Human Genome Project represented massive scientific collaboration from many participants to address a large research target simultaneously. The next step in the effort is to convert this data and knowledge into treatment regimens resulting in personalized treatments. The network of networks will be able to track the spread of epidemics and the effectiveness of treatments on an ongoing basis, provided confidentiality considerations do not make information inaccessible.

COMMUNITY: INCREASING EFFICIENCY FROM SPECIALIZATION YIELDS INCREASING INTERDEPENDENCE

The last 150 years of technological advance, combined with an economic system that rewards proper use of capital, has resulted in a material standard of living for U.S. citizens and first-world nations that is higher than that enjoyed by any people previously on Earth. The "natural selection" forced by the competitive economy and the accelerator of capital, attracted by the best opportunities, has caused best technological practices to be advanced, benefiting many millions of people. The vast majority of citizens in this society have access to food, healthcare, entertainment, and educational opportunities beyond the reach of kings two centuries ago.

Competition has caused a relentless journey toward efficiency and its counterpart: specialization. As we group together in ever-larger communities of specialized individuals, we become much more dependent on each other for the basic sustenance of our daily lives (food, fuel, etc). While our standard of living is far advanced over any ever known, reliance upon the productive output of others is also increasing. Such is the march of techonomics. The tangible sides of the organizational square — energy, computation, and communication — progress at accelerating rates linked inextricably to each other. What of the fourth side: community and its spirit? Will community life progress in the midst of superlative material comfort?

In his classic study of successful individuals, *Think and Grow Rich,* Napoleon Hill repeatedly observed that an individual's greatest success was never reached before a great failure.[10] Success had a price. Might that also be true for organizations and societies — that the times of greatest progress follow times of greatest challenge? Is there a collective will that contributes to the constitution of a people that is valuable to nurture, protect, and preserve, or are the citizens of a nation simple cogs in a scheme too large to influence? A superpower economy followed the Great Economic Depression. A superpower military followed the great struggle of World War II. The challenge from one man, John F. Kennedy, and fear of a competitor with a lead, the Soviet Union, urged a nation to send a man to the moon and back by a definitive date, and the nation responded.

Facing trials and overcoming them builds character in organizations as well as individuals. The free market global economy is creating many trials in this time of transition. It always has. The U.S. Civil War was a battle between the Agricultural and the Industrial Ages at the same time it was a battle for states' rights and freedom of slaves. How to maintain high material standards of living for the masses will be the key battle in the transition from the Industrial to the Virtual Age.

The enemies of healthy organizations, as well as of interdependent communities of working people, are excessive protectionism and entitlement. Certainly, societies have a moral obligation to protect their economic interests from being overpowered by businesses or foreign interests. But the entitlement mindset is a great destroyer of personal initiative and ultimately diminishes the ability of societies to effectively participate in the operation of global free markets and their benefits.

PROTECTIONISM

When governments intervene in an effort to protect segments of industry or population from the realities of the competitive economy, the result is always the same: descent into the abyss of mediocrity. The clearest evidence on an international level in my lifetime was the condition of the Soviet Union at the time of the fall of the Berlin Wall. The Soviet Union, with abundant natural resources and a vibrant people, had outwardly acted as a world superpower for 40 years. But on the inside, the government pursued a noncompetitive socialistic economic policy and could barely feed its populace. When the wall fell, none of the country's producers were strong or capable enough to assume a competitive role in the world economy. Fifteen years later, progress is still slow. Once destroyed, the infrastructure and will required to embrace competition is not easily cultivated.

The free market economy is the filter in techonomics providing natural selection of best technology practices and causing them to proliferate. When markets are contrived or unduly constrained, the long-term effects are analogous to creating a zoo in the biological world. Animals confined to the protective environment of a zoo for years or for generations lose the knowledge, cunning, and will to survive in the "wild." The zoo, where the weak and feeble are protected and allowed to procreate, fosters a sort of inverse natural selection.

Some years ago, red wolves were reintroduced into the Great Smoky Mountains National Park. These particular wolves had never lived in the wild. They were reintroduced into a protected area and fed with meat thrown over a tall wooden fence. After several weeks, the frequency of the feedings was reduced to encourage the wolves to begin hunting. Instead, the wolves spent most of their waking hours walking around the area looking up at the trees waiting for food to fall from the sky! Only a small number made the transition to freedom. The reintroduction program for the red wolf in the Smokies has been discontinued.

Protectionism in economic markets works the same way. The market cannot protect itself; quite the contrary. A market left to itself will reward best practices with economic success and punish poor practices with failure. Protectionist practices have to originate outside the system. The laws of economics govern the marketplace unless a powerful external influence dictates a modified set of standards. The outside influencer of major markets is commonly a national government — no other single entity has enough power over the market.

The U.S. government has, for example, used antitrust measures to protect the market. Whether it was the breakup of American Telephone and Telegraph (the Bell System), the paring down of International Business Machines, or the decoupling of Microsoft's key products, in the past when a company became (in the government's view) too successful and powerful, the government would act. In the age of multinational corporations holding little national allegiance, techonomics anticipates that this form of imposed corporate control will become much more difficult to impress. Wal-Mart is a good example. With its significant nation-sized revenues, Wal-Mart is capable of influencing the international balance of trade. Locally, Wal-Mart is such a powerful provider of jobs, employment taxes, and local sales taxes that it can frequently obtain assistance from local governments to procure desired land for new outlets. Few government officials will take a stand to weaken such a corporation, no matter how large it gets.

Will multinational corporations become so powerful they overshadow nations, or will nations, with tariffs, taxes, and antitrust regulations manage to control businesses? If Wal-Mart is any indicator, the multinationals are going to win if it comes to a test. Why? Techonomics. The multinational corporation in the age of virtual companies is no longer bound to location. It can move headquarters, distribution, manufacturing, research, and marketing to any location on Earth. If the government presses anticompetitive suits against a giant like Wal-Mart, they are hurting every consumer in the nation. Just about every voter is a consumer! Excessive business or employment taxes will have the same effect in the level playing field of this century. Just as U.S. companies often incorporate in Delaware before a public offering, due to tax considerations, multinationals will seek tax-sheltering countries if the tax and liability costs of headquartering in the U.S. are significantly more burdensome than in other nations.

When a nation chooses protectionism, which in turn encourages inefficient organizational practices, organizations lose the ability, will, or both, to compete. Economic protectionism is advanced by several methods: creating artificial trade barriers, providing unequal access to resources (land right-of-way, water access, broadcast frequencies, etc.) that result in an economic monopoly, or evaluating "competitive" bids on a basis other than economic merit.

If such practices ultimately result in weaker economic organizations, why does government perpetuate them? Most protectionist practices are an effort to gain political power or financial capital from those protected, or to maintain the status quo of past economic tradition (to maintain political influence with the general populace). However, they can also result from the unintended consequences of an action targeted at another market.

As observed with the ultimate demise of the economies of the Soviet Block nations, protectionism leads to economic failure. In the U.S., protection of the steel industry simply slowed its demise. In Japan, protection of the rice industry resulted in higher prices for rice and other grains.

Entitlement

What economic protectionism does to organizational competitiveness, cultural entitlement does to individual initiative and character development. Note this cycle: improved standards of living come from greater productivity. Greater productivity comes from greater specialization. Greater specialization produces greater interdependence, which maintains standards of living. Entitlement also runs a spiral course. ***As the entitlement cycle spirals, the expectation to receive more (standard of living) while doing less (specialization) leads to mass dependency (entitlement).***

With the establishment of the Great Society effort in the U.S. in the mid-1960s, the U.S. Government declared war on poverty. How has the war been going? Comparing income statistics from 1966 and 2003 provided by the U.S. Census Bureau in their historical poverty tables, it is clear that the percentage of people in the U.S. living below the poverty line has not significantly changed (1966: 14.7%, 2003: 12.5%) and has never been below 10%.[11] Even with the expenditure of billions of dollars to address the problem, the measurable economic results have been marginal at best.

The *desire* to produce is as important in the techonomic economy as the *ability* to produce. With computers and networks acting as leverage for the mind, anyone possessing the will to serve can find an opportunity to serve. Agricultural workers displaced by the industrial age did not have it as good. Industry was slow to develop and expand into the rural agricultural regions, leaving displaced workers with few options.

Computers and process have created a more level playing field for entry-level positions. But with welfare payments approaching minimum wage earning capacity, the motivation fueled by basic needs has diminished the will to work. Subsidized housing, subsidized meals, subsidized transportation, and subsidized healthcare challenge the thinking poor to consider the economic advantages of *not working*. The marginal economic benefits of minimum wage work are seen as negated by the constraining commitment of time and the costs (transportation, clothes, food) associated with work. The honor of self-sufficiency and pride of accomplishment has been erased in large segments of the community. **Techonomics anticipates that those cultures constraining or eliminating the entitlement mentality will be the cultures to dominate future markets.** Technology will assure that those who desire to be productive have the opportunity to be. Economics will assure that those willing to perform an honest day's work for an honest day's pay will be rewarded for doing so. Conversely, technology placed into the hands of one lacking a work ethic will never be optimally deployed, no matter how innovative. Global competition ultimately guarantees the demise of entitlement.

In this chapter, we have used techonomics to anticipate the key trends of our generation, the transition to the Virtual Age. To anticipate trends in the marketplace, techonomics assumes that the evolutionary powers of the free market are able to work. But when the free market is constrained, a techonomic analysis can also be used to study the detrimental trends in the market and offer corrective recommendations. The next chapter will review contemporary economic challenges of our time, seeking practical, workable techonomic remedies.

SUMMARY

TWENTY-FIRST-CENTURY TECHONOMICS

The transition to the Virtual Age from the Industrial Age is at hand. The key labor leverage in this transition is automation. The Industrial Age was about magnification of physical labor, whereas the Virtual Age is about magnification of mental labor without a physical presence.

FROM INNOVATION TO TREND

Many technology innovations die in the laboratory because of economic shortcomings (too expensive, too hard to use, etc.) When innovations are built upon the founding laws of techonomics, the economic certainty of reducing implementation costs and expanding customers aids success. Look for technology innovations that are based on two or more of the three twenty-first-century techonomic laws to find those with a high likelihood of mass adoption.

ENERGY: JOURNEY TO RENEWABLE ENERGY RESOURCES

The ongoing domestic struggle over energy — constrained by economy, environment, sustainability, regulation, technology, capitalism, and demand — is nearing a critical turning point. Techonomics observes that standard of living is closely tied to per capita energy consumption. To maintain living and environmental standards, domestic energy policies will have to be modified to allow economic and greenhouse-friendly energy conversion from nuclear fuels. Otherwise, we face diminished competitiveness and living standards.

COMPUTATION: ALL THINGS DIGITAL

Trends from the tangible to the intangible, atoms to bits, analog to digital, are all being fueled by Moore's Law. It is now cheaper to make a transistor than to print a single character of newsprint. The implications are that the world is moving from tangible to digital and digital to virtual — where most of our reality is perceived via our connection to digital devices.

COMMUNICATIONS: EXPANDING CONTROL AND INFLUENCE

We are in the first decade of ubiquitous, wireless, high-bandwidth, and global communications. Simultaneously, more personal access to information and more governmental oversight of individual actions exist today than ever before in the history of the world. The importance of ethical government — of, by, and for the people — cannot be overemphasized in light of the concentration of power provided by emerging technologies.

COMMUNITY: INCREASING EFFICIENCY FROM SPECIALIZATION YIELDS INCREASING INTERDEPENDENCE

Techonomics anticipates increasing efficiency that results from specialization and yields increasing interdependence. The key success determinant of the "community" going forward is the management of the entitlement mindset within an economy of plenty. Too high a level of entitlement will create a noncompetitive national economy watching other nations prosper as their technology-aided productivity expands. Too low a level of safety net will result in anger and chaos within a sizeable segment of the population.

KEY TERMS

renewable energy	acceptable risk
zero risk tolerance	digital omniscience
burdened labor rate	media convergence
hand-held convergence	organizational span of control
virtual company	automated bureaucracy
the Virtual Age	micro-multinational

QUESTIONS

1. Several emerging bellwether technologies are discussed in this chapter. Come up with one more of your choosing and explain how any of the first three Laws of Techonomics will help your bellwether technology become mainstream in the years ahead.

Emerging Techonomic Trends

2. Energy production and generation is a key to economic growth, yet the U.S. has not licensed a new electricity-generating plant fueled by nuclear power in 30 years or a new petroleum refinery in 27 years. Given that these facilities can take from 5 to 15 years to design and construct, the U.S. is in the middle of a 50-year gap in new energy production facilities for these fuels even if we act with boldness today. What does this lack of resource development mean for the future U.S. standard of living? For the balance of trade?
3. The hydrogen economy is touted as the future of clean energy production. When hydrogen "combusts," its byproduct is water (no greenhouse gases). The common way to obtain hydrogen is via electrolysis of water, requiring electricity. Emerging methods to produce hydrogen include high-temperature decomposition of water (> ~2000°C) using solar or nuclear as the energy source for the conversion. Certain microbes are also being studied for their characteristics to offgas hydrogen. Survey the Web to (a) identify viable methods of hydrogen generation, and (b) develop a personal estimate as to how many years it will take for your viable solution to make it to the marketplace. Would you fund research into this endeavor?
4. Ethanol has been put forward as a clean-burning fuel for our future transportable energy needs. I disagree, because its combustion still produces carbon dioxide. Proponents of ethanol say that it burns much cleaner than gasoline, and the demand it would create for agricultural products would stimulate the domestic farm economy. Create an energy-consumed and energy-produced balance for energy obtained from corn-based ethanol. What are your conclusions? Can you identify any technology breakthrough that would change this deficit in the energy conversion cycle?
5. More electric power is generated from nuclear power today than was consumed by the entire planet in 1961, yet we in the U.S. continue to constrain the use of this energy form. What steps in technology, economy, and regulation are needed to move this industry forward? Should those steps be considered, or what alternatives do you recommend to meet future energy needs?
6. Computation and communication advances have accelerated simultaneously the last decade. As a result, the cost of a transoceanic telephone call has fallen almost two orders of magnitude. This has diminished the "communication" wage differential for information-intensive labor (telemarketing, accounting, drafting, software development, etc.). Simultaneous with this techonomic trend has been the rising cost of healthcare insurance in the U.S. Comment on the shift to a global labor force for information-age work if the transoceanic phone costs have gone from $1.00/minute to $0.05/minute while the cost of domestic healthcare insurance has gone from $1.00/hour to $5.00/hour. Can you offer any techonomic predictions for globalization of the information technology workforce given these ongoing trends?

REFERENCES

1. Svoboda, E., UC scientist says ethanol uses more energy that it makes, *San Francisco Chronicle,* Monday, June 27, 2005, http://www.sfgate.com/cgi-bin/article.cgi?f=/c/a/2005/06/27/MNG1VDF6EM1.DTL&hw=ethanol&sn=001&sc=1000.
2. Segelken, R., CU scientist terms corn-based ethanol "subsidized food burning," *Cornell Chronicle,* 33, 6, 2001, http://www.news.cornell.edu//Chronicle/01/8.23.01/Pimental-ethanol.html.
3. Wind, T., Windustry WindProject Calculator, Windustry, 2003, http://www.windustry.com/calculator/default.htm.
4. Bellis, M., History of Photovoltaic Cells, Inventors, About, About Inc., 2005, http://inventors.about.com/od/pstartinventions/a/Photovoltics.htm.
5. Nanosolar, Inc., Understanding cost/performance, Nanosolar, Inc, 2005, http://www.nanosolar.com/understandingcost.htm.
6. Negroponte, N., *Being Digital,* 1st ed., Knopf Publishing, New York, 1996.
7. Sennett, R., *The Culture of the New Capitalism,* Yale University Press, New Haven, CT, 2006.
8. Sellner, Manfred B., *Babal or Behemoth: Language Trends in Asia,* (review), International Institute for Asian Studies, IIAS Newsletter 34, 29, 2003, http://www.iias.nl/iiasn/july04/bb.pdf.
9. Smolan, R., *One Digital Day: How the Microchip is Changing Our World,* Crown Business, New York, 1998.
10. Hill, N. and Cornwell, R., *Think and Grow Rich,* original version, restored and revised, Aventine Press, San Diego, 2004.
11. U.S. Census Bureau, Historical Poverty Tables, U.S. Census Bureau, 2004, http://www.census.gov/hhes/poverty/histpov/hstpov2.html.

Section 4

Postindustrial Challenges and Techonomic Answers

8 Techonomic Market Crises and Recommendations

> If you could kick the person in the pants responsible for most of your trouble, you wouldn't sit for a month.
>
> — Teddy Roosevelt

TECHONOMICS NATURAL SELECTION MECHANISM: COMPETITION

The relentless march of knowledge in applied form (technology) has always been the force behind human productivity. In product after product, service after service, endeavor after endeavor, and market after market, continuous improvement in technology and its expanding deployment have resulted in more output with less human labor and material resource. Techonomic progress has reshaped the fundamental productivity of organizations on all levels throughout history.

In view of the applicability of techonomics to so many fields, crossing time and national boundaries, and with the rate of techonomic change accelerating, could there possibly be any current markets where techonomic forces are not working? Can a market defy techonomic trends? Unfortunately, some markets and endeavors not only defy but reverse them. They can do this only when a powerful external force is exerted and maintained on the marketplace, however. And the price is high. *As the technology gets better, the economic cost to provide endeavors is increasing, because these organizations are failing to become more productive.*

Under normal circumstances, the techonomic theory of organizational evolution would anticipate that the competitive market, the natural selection mechanism, would eliminate negative productivity trends. Remember, each key element in Darwin's theory of organic evolution has an analogous element in techonomic theory. In the competitive economic arena, it is techonomic intelligence that gives some organizations the edge over others and results in their survival and growth, just as in the natural environment, it is some favorable characteristic — better eyesight, coloration, speed, etc. — that gives an animal the competitive edge and "selects" it for survival, growth, and reproductive success.

Competition in an open market, where techonomic forces can operate freely, is the selection mechanism that favors productive organizations and promotes their health and growth. Fundamentally, techonomic theory posits that technology advance results in productivity gain, thereby providing greater output for less cost. This theory also anticipates an inverse result when free economic competition is artificially constrained.

In three major U.S. markets — energy, healthcare, and education — techonomic trends are operating in reverse, resulting in an economic decline headed to crisis. The missing element in each of these markets is competition. Each of these troubled markets exhibits a different way of diminishing the corrective force of competitiveness, but the external force perpetuating the artificial market originates from the same source — government:

- **The energy market:** Domestic competition from nearly all energy sources has been stymied by unilaterally expensive environmental regulations. These regulations render domestic exploration and production economically unjustifiable relative to international sources.
- **The healthcare market:** By removal of personal responsibility within a multilayer system of insurers, providers, litigators, beneficiaries, and suppliers, competition has been diminished. The motivation for efficiency, beginning with the individual, has been eliminated because "someone else" pays the bills.
- **The educational system:** Monopoly leads to waste and inefficiency. The public education system is a monopoly with a firm agenda to maintain a perception of adequacy.

In this chapter we shall look at the negative trends in each of these markets, and we shall consider specific suggestions for reintroducing competition into these key markets. Due to the size and importance of these three markets, strength within them is actually vital for sustaining the competitiveness and health of our remaining economy. However, we shall no doubt have to suffer the ill effects of these negative trends for a while longer before the will of the American people requires its government to free these markets.

Since techonomic trends are active in the global economy, performance of these U.S. markets must be viewed in terms of contemporary world markets. Techonomics predicts that U.S. trends in these markets are very likely to be corrected in reaction to a reduced economic living standard in our country relative to the world. However, if the nation does not act decisively to address the major noncompetitive markets producing these combined trends, then **more efficient and effective nations tempered by the competitive fire of the international marketplace will economically prevail.**

In addition to competition, the free market attracts capital, bringing resources to productive causes. Capital formation, the act of acquiring monetary resources for the execution of a productive endeavor, is increasingly more important as technology advances. The acquisition of capital for an endeavor demands a marketplace wherein a reasonable return is possible (not guaranteed, but at least possible). It has been aptly said that entrepreneurs are not risk takers; they are calculated risk takers. If an opportunity is going to attract the capital it needs to grow, there must be a reasonable governmental, economic, and cultural system in place that respects the

Techonomic Market Crises and Recommendations 159

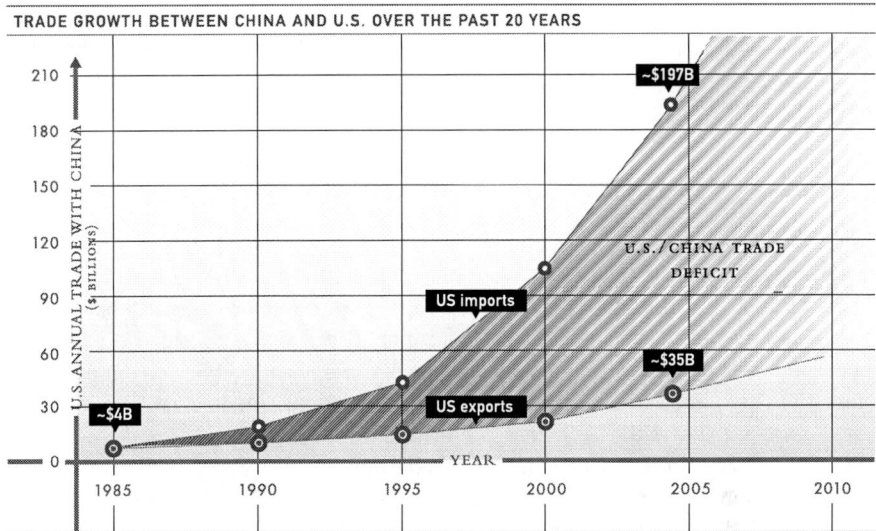

FIGURE 8.1 Growth in trade between the U.S. and China over the past 20 years.

importance of capital. Investments must have the opportunity to bear fruit. If such conditions do not exist, then elevated capital risk demands payback rates and shortened time frames that reduce or eliminate the financial foundation needed to proceed.

The current situation in China is a clear example. In recent years, China has systematically opened up trade and free market practices on its borders. Figure 8.1 shows the growth in trade between the U.S. and China over the past 20 years.[1] *Capital formation for China has had to come from revenues generated from product sales and from corporate partnerships with foreign firms willing to bear the risk of unknown capital markets and government policies.* Conversations with leading international equities brokers have confirmed my conviction that investment in equities listed on Chinese exchanges remains a very risky endeavor. The high risk is there mostly because of the uncharted performance of their equities regulatory oversight system. China's transformation will be slower as a result, but it will still progress because of free trade and free markets around the world seeking low-cost goods, wherever they can be found.

The techonomic productivity (output/cost) of manufacturing in China is increasing rapidly as China embraces the competitive world marketplace. By contrast, the techonomic productivity of manufacture in the U.S. is on the decline because we are burdened with decreasing performance in the three influential markets: energy, healthcare, and education. *Techonomics* has already considered aspects of these markets both historically and recently, but this chapter offers analysis to pinpoint the means by which marketplace competition was removed and to suggest corrective measures. ***To restore competition, it is essential to understand the means by which it has been constrained.***

Application of techonomics to markets in crisis is extremely important; the economic burden of national failure in these three markets could be large enough to cripple the entire private sector in the U.S. Without change, our trade deficit, budget deficit, and tangible productive capacity will slowly, steadily, ultimately evaporate. This will leave the U.S. economically — and in the end, militarily — subservient.

ENERGY: ECONOMIC REASON OR RUIN

Fossil fuel was the energy of choice for the industrial age. Coal, petroleum, and natural gas, hidden for the ages, were extracted and released over a period of 200 years to fuel the most rapid economic progress in human history. Debates about fossil fuel and the environment (CO_2 and global warming), and the size of world petroleum reserves, have raged for the past 30 years while U.S. consumption and trade deficits related to energy use have grown relatively unchecked.

Figure 8.2 shows the U.S. energy consumption from nondomestic sources over the last 40 years in terms of barrels and constant dollars. Today, over 60% of the oil we consume is imported. Faced with a similar, though less vulnerable, situation in 1979 with the Arab Oil Embargo, U.S. imports of foreign oil dropped significantly over a period of three years (Figure 8.2). Regulations on automobile fuel efficiency were implemented. Where possible, economically competitive fossil fuel sources like coal and natural gas were substituted for the petroleum commodity, which was increasing in price at a rate beyond market tolerance. And nuclear power at the time remained a viable electrical energy producer. Over a period of a few years, the market adjusted as would be expected.

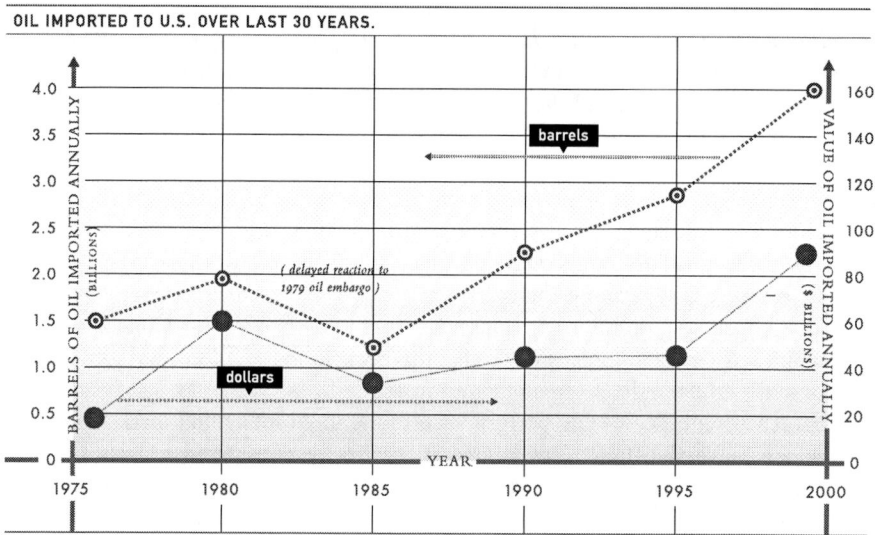

FIGURE 8.2 U.S. importation of crude oil over the last 30 years in terms of barrels and constant dollars.

Fast forward to the twenty-first century. According to the U.S. Department of Labor, the combined expenditure of energy for transportation and home use exceeds expenditure on food of all types.[2] Bottom line: we spend more on energy than on food. And the gap is widening, since the nation has shown no will for reducing its imports, even as the vulnerability of major refineries has been exposed by nature. The lack of will to solve this problem does not go unnoticed by those who profit from our ineptitude. They see a competitive world in which they hold precious resources, and a national buyer with no alternatives for energy independence.

The Oil Producing and Exporting Countries (OPEC) learned some valuable economic lessons from the response to embargos in the late 1970s. *A monopoly can control supplies and prices only to a certain degree before a resourceful customer overcomes switching costs and inconvenience to seek alternatives.* But the more difficult the substitution of alternative fuels (i.e., the higher the switching costs), the greater the economic pain before alternative sources will be sought. Our culture has become dependent on foreign fossil energy, much like some individuals become dependent on addictive drugs.

Only severe economic pain will change the nation's response, but the hands at the oil controls have learned not to be too severe, lest we awaken from our folly. Remember the old adage about boiling a frog? You do not get the pot of water boiling hot and then throw the frog in; he will immediately jump out. You put your frog into lukewarm water and then very slowly turn up the heat. The frog never perceives his danger. The water eventually boils and cooks the frog.

Incremental price increases of a dollar or two a barrel ($0.05 a gallon at the pump) once a quarter or so, over many years, slowly boils the oil consumer. Add environmental activism prohibiting domestic development of energy resources, and you get to a point where the consumer is entirely cooked. *Because of environmental constraints on petroleum exploration, petroleum refining, coal production, and nuclear generation, the U.S. is much more limited today than in the late 1970s in its ability to respond to any shortfall in petroleum.* The immediate escalation of gasoline prices in the week following Hurricane Katrina is practical evidence of this techonomic reality.

The NIMBY (Not In My Back Yard) approach has not only constrained oil exploration and production, it has limited the development of virtually every other alternative. The economics of coal production and clean combustion have been detrimentally impacted to the point that large quantities of coal from the east coast cannot be considered for use without extensive sulfur removal processing on both the raw feed material and the exhaust emissions. The economics of nuclear plant design and construction have been rendered intractable by a licensing process fraught with detours and modifications. The remote possibility of a catastrophic event makes the acquisition of liability insurance economically infeasible without government intervention. As a result, rather than benefiting from the presence of these energy-producing capabilities in this country, where they can be reasonably regulated, we are forced to import ever-greater units of energy from abroad — much to our economic detriment. In this global village called Earth, airborne mercury produced in China comes to rest in the Appalachians. Carbon dioxide generated from unregulated deforestation in Brazil adds to global warming worldwide. The U.S. invites economic ruin the longer it insists on unilateral environmental regulation eliminating reasonable economic exploration and production of its own energy resources. ***No sovereign nation would agree to unilateral military disarmament, yet the U.S. has acceded to the energy equivalent of unilateral economic disarmament.***

In earlier sections, *Techonomics* reviewed energy sources available to address our needs if the marketplace is allowed to seek rational, balanced technical solutions: nuclear for centralized electric power generation, solar for distributed electric power generation, and hybrid gas-electric to drastically reduce foreign fossil fuel dependence. Policies promoting other sources are motivated by special interests (ethanol from plants), flawed economics (wind and wave power), or flawed technology understanding (hydrogen cycle). Techonomic trends through the ages reveal that general standard of living is linked to per capita energy consumption. To progress, we need abundant, clean, inexpensive energy.

Widespread, decentralized solar energy use will happen when solar technology cost reductions materialize. It is a matter of economics and system dependability. Hybrid vehicles are a near-term answer to the extension of precious hydrocarbon fuels, and they operate more cleanly and efficiently. But the most significant opportunity for a near-term improvement in the domestic energy outlook is to focus societal resources on effective deployment of nuclear power. Without nuclear, we will continue to foul the air and import ever-increasing tons of hydrocarbons. But with nuclear power, abundant, clean electrical energy to meet growing needs is a realistic possibility. The technical problems are tractable and few. All the major stumbling blocks rest within a fearful culture that does not embrace the concept of acceptable risk within the greater concept of economic viability.

The question remains: Will the U.S. change regulations to encourage capital formation for construction and operation of nuclear plans? The following steps must be taken before commercial capital will become available for nuclear facilities:

1. Unify the regulatory system. In the past, multiple licenses for design, construction, and operation made the system a labyrinth of changing conditions. The process must be unitized and the regulations fixed once a project is approved.
2. Cap liabilities and insurance premiums. Infinite risk with infinitesimal probability results in an undefined financial result. Insurance companies and boards of directors do not like undefined results. If nuclear energy is to be a real option for the future, limits of liability and economically viable liability premiums will have to be established. This is an example of an early-stage industry where the government must act in its own best interest to encourage, rather than discourage, capital formation. Otherwise, there will not only be no capital formation, but ultimately there will be no energy.
3. Establish a waste disposal process. Nuclear wastes must be properly stored and monitored. The technology and processes exist, but the will to initiate operations of the storage facility has been lacking.
4. Standardize plant design. First-generation nuclear plants differed in design for every location. A standard design, easily replicated with standardized

procedures and replacement parts, will greatly improve the economics of design, construction, and operation.

As long as the "calculated risk" cannot be calculated, the nuclear industry will remain dormant. When government policy shifts to a long-term, positive support of nuclear power, these fundamental changes will revive the industry. While techonomics does not favor subsidies, it does favor removal of societal roadblocks and placement of realistic limits on liabilities. Without these steps, nuclear power will remain the cleanest, most economical alternative not to be used. Only increased economic pain for the masses will force the change to a realistic policy. Such pain is on the horizon. Unfortunately, even under the most positive circumstances, it will take more than a decade to go from project commitment to electrical generation. If the geopolitical climate in the world's oil-producing region were stable, there might not be such cause for concern and need for action. If the entire productive economy of the private sector did not run on oil, there might be much less concern for action.

Field: Energy.
Missing economic natural selector: Domestic vs. international regulation constrains domestic energy production while fouling the global environment.
How imposed: Expensive environmental and circuitous licensing regulations.
Anticipated techonomic result: Uncorrected, long-term reduction in domestic standard of living will stem from reduced economic competitiveness due to high energy costs. Domestic environmental regulations leave the international air unguarded, yielding more global pollution.
Techonomic opportunities: Encourage nuclear by streamlining licensing and standardizing plant design, focus on cost reduction efforts for solar voltaic collectors, and encourage electric-gas hybrids to stretch efficiency and reduce pollution.

HEALTHCARE: INVERTED TECHONOMICS AND ITS IMPLICATIONS

A simple techonomic metric for healthcare, the lifecycle healthcare expenditure per person, reveals a techonomic trend toward diminishing productivity; as technology has improved, the life cycle cost of healthcare has increased — significantly. The U.S. healthcare system inverts techonomics by the combined impact of more medical procedures, longer average lifespan, and an economic structure that shifts responsibility from the individual recipient of services to the collective society represented by various inefficient intermediaries. Encapsulate the healthcare industry within a broader society consuming processed and fast foods while leading a more sedentary lifestyle, and the resulting trends point toward an economic reckoning.

Figure 8.3 shows the change in the techonomic metric for healthcare in the U.S. by multiplying life expectancy at birth times the annual per capita healthcare costs.[3-4] This is the techonomic metric developed in Chapter 4, and the results over the last

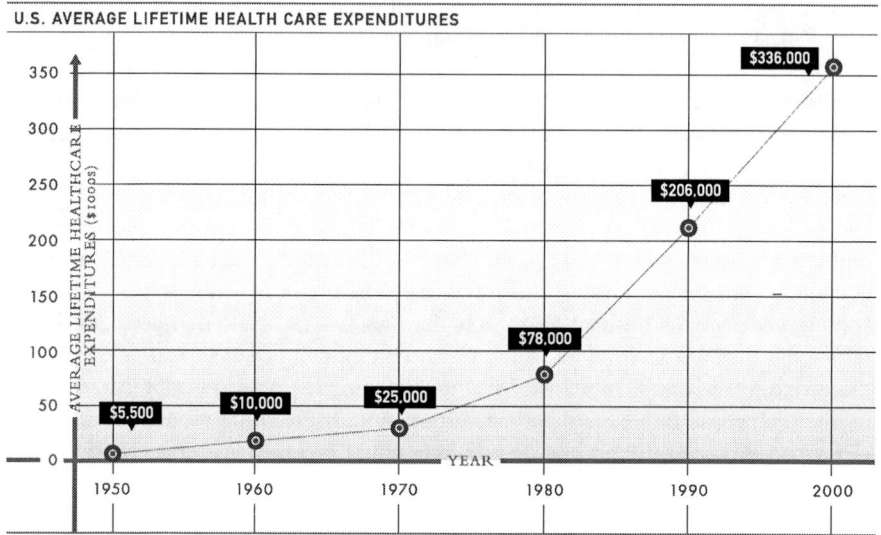

FIGURE 8.3 Change in the techonomic metric for healthcare in the U.S. by multiplying life expectancy at birth times the annual per capita healthcare costs.

40 years show an alarming techonomic trend: a significant part of the economy spiraling out of control. Dollars spent on healthcare over a lifetime in the U.S. have increased from $10,500 in 1960 to over $355,000 in 2000.

Healthcare costs now exceed *$15 dollars a day* for every man, woman and child in the country. These costs are rising at twice the rate of general inflation. Another way to look at the significance of the healthcare crisis is to observe healthcare expenditures as a share of the total gross domestic product. The portion of the Gross Domestic Product (GDP) devoted to healthcare in 1960 was 5.1%. By 2003, that same portion was 15.3%.[5] This trend is not sustainable beyond the current generation.

This significant trend leads to only three possibilities or some combination. As a nation, we are either (1) producing less in the whole economy, resulting in a GDP drop, (2) requiring greater healthcare system access, or (3) strapped with a healthcare structure that is not sustainable.

The first possibility, less productivity, is ruled out as the source of dramatic change, because the GDP has continually grown through the years. Only for short periods in the midst of recession has the GDP held steady or decreased slightly. As a nation, we are requiring more healthcare access. As technology progresses, we are developing new procedures and medications. Incremental increases in life expectancy have occurred over the past 40 years (1960, 69.9 years; 2003, 77.6 years). Better nutrition, new surgical procedures, and advances in pharmaceuticals have all contributed to this 10% increase in average anticipated lifespan. But the improvement has come at a cost: a 60 *times* increase in healthcare spending between 1960 and 2003. A techonomic analysis anticipates that ***the economy cannot indefinitely sustain such incremental improvements at such great expense.*** A significant increase

Techonomic Market Crises and Recommendations 165

in cost for a marginal return reveals a faulty market structure. Techonomics is not working in healthcare, but why?

There are many reasons why the healthcare system displays declining techonomic trends. They include:

- *Lack of personal responsibility:* The individual covered by an employer or government program has little economic incentive to seek lower-cost provision or to take care of their personal health maintenance. There are no economic rewards for good health choices, good shopping of services, or avoidance of unnecessary services. When was the last time *you* got an estimate for medical services from more than one provider before selecting a hospital for a treatment? Options from the list of "approved" hospitals are usually selected based on convenience and where your doctor practices, not based on any economic, competitive basis.
- *Quasi-monopolistic intermediary/multilayer system inefficiency:* Powerful insurance companies have little motivation to minimize costs other than massive buying power with the providers. If they experience a bad year where claims are higher than expected, they simply raise their rates the next year to cover any previous shortfall and forward-looking profits. Primarily, it is employers who feel these rates changes. There are very limited numbers of competing alternatives in the marketplace, and switching procedures are arduous. This market approach is contrived for the benefit of the insurers, and the power of their lobby on the state level will cause it to remain so.
- *Unlisted pricing policies for service providers:* Insurance providers use collective bargaining methods to obtain most favorable pricing from hospitals for services rendered. Individuals cannot obtain a price quote for hospital services in order to compare between providers, and when they are billed, they carry substantial overhead costs for nonpaying customers.
- *Economic impact of unlimited liability:* The impact of "ambulance chasers" has caused malpractice insurance rates to climb dramatically. Some areas of practice, like gynecology, have been greatly affected by rising malpractice insurance premiums. These high premiums, resulting from a trend of more and higher settlements in malpractice cases, are simply adding to the cost of doing business for the professionals in the field and, in some cases, forcing doctors to move or leave their practices.

There are other key contributors as well: more prescriptions, greater cost per prescription, inefficient hospital practices, high salaries, Medicare fraud and inefficiencies, etc. **Left unchecked, healthcare expenditure relative to GDP will exceed 20% before the end of the decade.** The aging population, increasing labor costs, and disproportionate end-of-life healthcare expenditures will keep this trend growing. The national economic system will not be able to support this burden and remain competitive internationally. Leaders of Starbucks lament that they now spend more on healthcare insurance benefits than on the material for their products! The cost of the healthcare benefit frequently paid by companies is now a significant contributor to labor cost contained within each product. Healthcare costs represented the largest component of personal consumption spending, at 16.7% in 2000. Healthcare spending is followed by food (14.1%), housing (13.6%), transportation (11.8%), and household operations (10.7%).[6]

Rising insurance costs will force an ever-increasing number of employers to withdraw from the current system in order to remain competitive. "The cost of family health insurance is rapidly approaching the gross earnings of a full-time minimum wage worker," said Drew Altman, president and CEO of the Kaiser Family Foundation. According to the 2004 Annual Employer Health Benefits Survey, premiums reached an average of $9,950 annually for family coverage ($829 per month) and $3,695 ($308 per month) for single coverage. Family premiums for PPOs, which cover most workers, rose to $10,217 annually ($851 per month) in 2004, up significantly from $9,317 annually ($776 per month) in 2003. According to Jon Gabel, vice president for Health Systems Studies at the Health Research and Educational Trust, "Since 2000, the cost of health insurance has risen 59%, while workers wages have increased only 12%. Since 2001, employee contributions increased 57% for single coverage and 49% for family coverage, while workers' wages have increased only 12%. This is why fewer small employers are offering coverage, and why fewer workers are taking-up coverage."[7] When workers are left without healthcare insurance, more individuals are placed in government programs like Medicaid or are simply not covered. These workers may be able to afford coverage, but the insurance system is set up for groups, not individuals, and health insurance is difficult for individuals to get.

Healthcare is precious and dear to our lives, particularly when it is our personal healthcare. But when the government assumes the financial responsibility for our health, it also makes the rules. As individuals, avoiding premature death is a win-at-all-cost battle. Currently, the system makes financial gain while attempting to save life at any cost. Individuals desire it, the government pays, hospitals generate revenues, and insurance companies pass on the costs to business via rate hikes. Simple system, but it is unsustainable in a global, competitive market. Currently, 1% of the population consumes over 22% of annual health spending in the U.S. At the same time, 50% of the population consumes less than 4% of the annual spending.[8]

We have a healthcare system that spares no expense to deal with litigation, advancing technology that works miracles, and medical plans that (after small personal deductibles are paid) allow families and individuals to pursue extended life by any technical means. The patients do not pay; others do. The techonomic per capita lifetime costs of healthcare now tops $330,000 per person. And with 45 million citizens not on any insurance plan, the trends of higher costs and greater government subsidies are significant and growing rapidly.

As a nation, we stand at a crossroads: socialized healthcare or free market healthcare. The economics of the situation will force changes. The system is unable to bear the economic weight of being all things to all people, particularly when those people are not making healthy decisions in their living patterns. ***Technology advances are prolonging life in a myriad of new ways, but we have not obtained the discernment as a culture to apply the law of diminishing returns as it applies*** *to the use of life-extending technology.* As a result, by the most conservative estimates, we spend, on average, 30% of the lifetime costs for a person in their last year of life (other estimates put that number at over 75%).

Moral, ethical, and technological questions outpace answers. What is a reasonable quality of life? How much should the government be expected to provide toward sustaining an individual's health? Are there international models where healthcare is provided universally, the quality is acceptable, and business has been able to remain globally competitive? Who is responsible for deciding about, and providing, your healthcare? How can our society reduce the entitlement mentality disproportionately driving demand for services? There are answers, but no easy ones. As with the energy dilemma, significant political will is required to set this market on a new course. Such will seldom surfaces prior to a significant crisis.

Field: Healthcare.
Missing economic natural selector: Personal responsibility and competitive pricing options from providers.
How imposed: System manipulated to advantage by powerful intermediaries (insurance providers, political gain, political lobbies).
Anticipated techonomic result: Large, hidden insurance benefit expenses embedded in labor costs are compromising business competitiveness. The current system will slowly implode as businesses limit or eliminate coverage benefits.
Techonomic predictions: Distinct choice between socialized medicine or personal responsibility is emerging. A dual-provider system is anticipated, where elective procedures are available for a price from private providers. Medical savings accounts combined with catastrophic insurance coverage offer a path to direct consumer responsibility and choice in the marketplace. Only personal responsibility combined with clear, competitive provider pricing options will promote improved techonomic trends.

EDUCATION: TECHONOMICS OF MONOPOLY

The problem drives the learning.

— Thomas Pate, Vice-president for training, Cracker Barrel

Don't be satisfied with mediocrity.

— W J Julian, Retired Director, Pride of the Southland Band, University of Tennessee

We have two distinct categories of education in our country: K to 12 and higher education. These educational systems face a world changing faster than they are adapting. Each system faces techonomic challenges for completely different reasons, and they are discussed separately.

K TO 12 EDUCATION, PUBLIC AND PRIVATE

Michael Hodges, in his series The Grandfather Reports, provides an insightful review of current performance trends in K to 12 education. Using a techonomic metric called U.S. Educational Productivity (SAT test score/annual $ per student normalized to 1960), he tracks the productivity decline of the U.S. educational system. Figure 8.4 shows the U.S. Education Productivity Index from 1960 to 1994.[9]

In 1995, according to Hodges, the SAT test scoring process was totally revamped so as to not be comparable to prior years' performance! **In constant dollar terms, this metric reveals a 71% decline in educational productivity over a 35-year period.** A rapid rise in spending is accompanied by declining results. This is techonomics in reverse.

Mathematics is at the heart of technological understanding, whether the scientific field is physics, chemistry, engineering, biology, etc. Proficiency in mathematics is required for success in technology. Unlike other disciplines where the important facts are subject to judgment and opinion, mathematics is an exact, rigid discipline with progress based on layers of mastery. Since technology is the driver of change within organizations (and nations), and mathematical education is at the heart of technology, a good way to evaluate international educational competitiveness is to compare mathematical performance of students.

The Organization for Economic Cooperation and Development–Program for International Student Assessment (OECD-PISA) has made an internationally recognized effort to compare students' skills in mathematics, science, and literacy. Over a quarter million 15-year-old students, scientifically sampled from 41 nations, participated in the assessment. U.S. students placed 24th of 29 countries participating in the OEDC study — far below the average in mathematics performance. U.S.

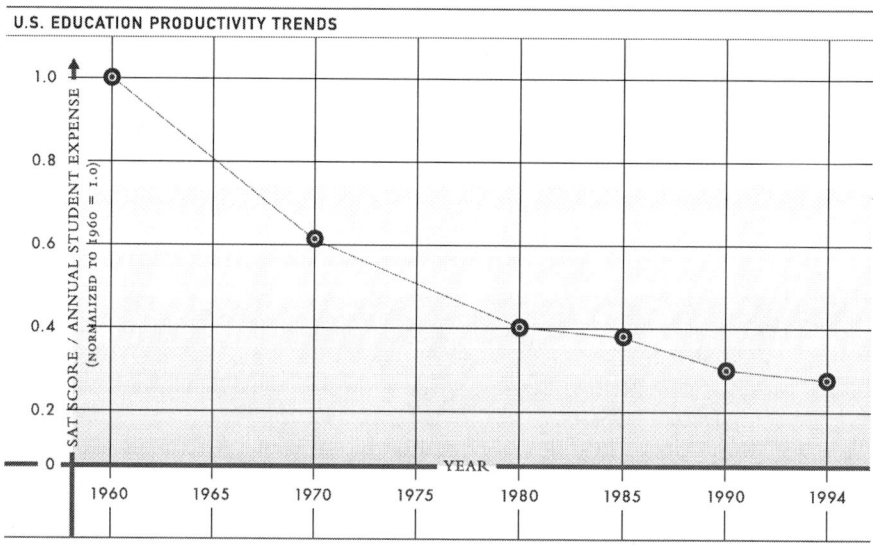

FIGURE 8.4 U.S. Education Productivity Index from 1960 to 1994.

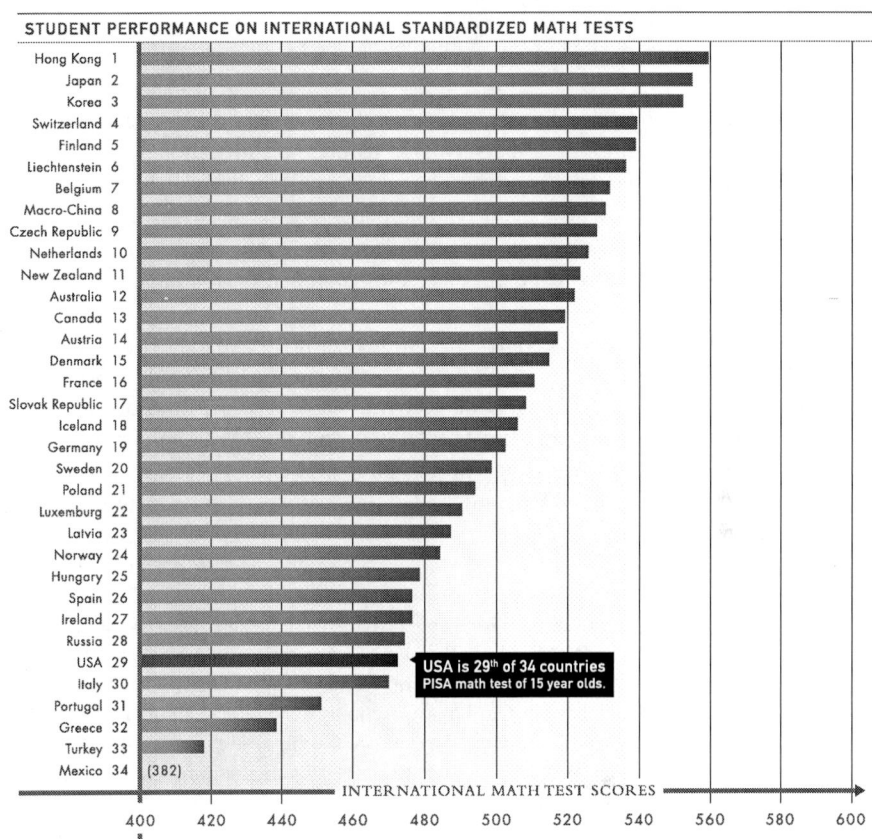

FIGURE 8.5 Average country scores of all students in the mathematics portion of the 2003 OEDC-PISA tests.

student surpassed only the students in Turkey, Portugal, Mexico, Italy, and Greece. Even the top 5% of U.S. performers fared no better against their counterparts in the world. Figure 8.5 shows the average country scores of all students in the mathematics portion of the 2003 OEDC-PISA tests.[10] The OECD report cited the U.S. mathematics score as "significantly below average."

This is one test in one subject, but the trend is clear in numerous international tests of high school math and science skills; the knowledge and performance of U.S. students is trailing the industrialized, and even many third-world, nations. What is bogging down the operation of techonomics in K to 12 education in the U.S.? Monopoly. If monopoly eliminates competition as the "natural selection" motivator for improvement, then the resulting system has no basis for progress.

Over the last 40 years, the public education system has embraced an ever-expanding socialization role — to the detriment of education. Busing eliminated the neighborhood school and, with it, much parental involvement. Feeding programs and extended care at schools reduced parental responsibility for the process of raising children as the government assumed an ever-expanding role as social agent. The layers of educational bureaucracy and inefficiency have grown so large that, as of 2001, more educational employees serve as administration and staff than as teachers.[11]

In the process, the school system seriously compromised its core mission — education — while attempting to be surrogate parents for the nation's children. But the system has not even been good at surrogate parenthood; it has been more like a permissive babysitter than a good parent. It has not only failed to prepare our youth to compete internationally in a global economy, it has also undermined core family values while taking over an increasing number of fundamental parental responsibilities.

Two fundamental flaws pervade the national structure of our K to 12 public education system: it is free and it is a monopoly. Actually, education is not free at all. In my own county, K to 12 education accounts for over 65% of the annual budget — more than sheriff, parks, and fire protection combined. Yet, the consumer perceives it as free, because there is no choice about where to participate, and participation is mandatory. Napoleon Hill anticipated the decline of our educational system when he wrote:

> We have in this country the greatest public school system in the world. We have invested fabulous sums for fine buildings. We have provided convenient transportation for children living in rural and other areas. But there is one astounding weakness to this marvelous system — it is free! One of the strange things about human beings is that they value only that which has a price. The free schools of America and the free public libraries do not impress people because they are free (or appear to be so). This is the major reason why so many people find it necessary to acquire additional training after they quit school and go to work.[12]

Hill had studied the successful leaders of his generation for 27 years, learning about human nature, success, and unleashing potential. Even in the midst of the nation's Great Depression, he observed that people do not value things that are free.

One simple and dramatic K to 12 education system change — vouchers — would bring back the force of competition. The simple act of placing educational purchasing power in the hands of the parent/student would remove the mindset of "free" and instigate competition at the same time. The resulting reshaping of the system would be traumatic, massive, and significantly positive for all people in our country, not just the affluent capable of buying their way out of the system.

Currently, we have dual K to 12 educational systems: private and public. Many affluent citizens choose the private system or locate their residence specifically to take advantage of an exemplary public school program. The less fortunate are relegated to a system producing third-world results. But vouchers would force choice and competition. The socioeconomic customers — students and parents — would have the freedom to participate in the selection of the best environment for their student. This empowerment is absolutely required by techonomic analysis to reverse the trends that are hampering our educational system. Any method that provides universal competition, and hence empowerment of consumers, would be positive for this market. But vouchers are the most direct approach to such a result.

A reasonable voucher program would provide 75% of expenses for the average local school pupil, paid to the parent(s) to use at any school, while sending 25% to the system. The system funding would support baseline operations even for students not enrolled. The result would be higher per-student revenue to the public schools. It would also force public schools to vie for students with the private sector and other public schools. The public system would shrink as excellent private schools expanded, but the per-student funding in the public system would actually increase as a result of the 25% payment for every child. This would allow the continuance of government programs for special needs that are not met elsewhere. Best practices and successful operations would expand, while failed schools would be forced to improve or cease to exist: educational evolution for K to 12.

Field: Education, K to 12.
Missing economic natural selector: Customer choice, eliminated by the monopoly provider.
How imposed: The public educational system, supported by teacher's unions and special interests, has perpetuated mediocrity. Employees, not customers, control the system.
Anticipated techonomic result: A continuing decline in student performance relative to international students will occur until "natural selection" of competition from choice becomes the norm for the system. Dual education systems (public and private) will continue to diverge in student performance.
Techonomic recommendation: Vouchers, the most straightforward method for infusing competition in the shortest period of time.

HIGHER EDUCATION

Our system of higher education, since it exists in a market full of options and competition, differs from K to 12 education. Higher education is faced with a different set of challenges associated with technological advance, diminishing public subsidies, rising costs, and issues of relevance in a fast-changing era. According to Dr. Harville Eaton, President of Cumberland University, the higher-education system in the U.S. has experienced three distinct phases:

1. **Colonial Phase**: Small elite colleges and many academies training teachers, ministers, and community leaders.
2. **Land Grant Phase**: The Morrill Act of 1862 was established to create higher education opportunities in each state to promote agricultural and mechanical instruction for all social classes. During this phase, the nation made great strides in agricultural productivity, transitioning from the agricultural to the industrial age.
3. **Post-WWII Military Research Phase**: In this phase, the higher education system began to provide research for the U.S. government-military-industrial complex. Research funding related to military needs became a focal point of funding for many institutions.

Throughout these phases, the teaching methods used to educate students remained virtually untouched. Students came to a classroom, heard lectures from a teacher, read books, and were tested on the material they were expected to understand.

We are now entering the age of virtual education. The three fundamental techonomic laws are active in reshaping higher education via distance learning: ubiquitous computation, ubiquitous communications, and increasing productivity gains. Combine a transportation system expanding geographic considerations for student recruiting, and the range of possible choices is greater than ever before. Because of competition from virtual education, competition for students is keener among those institutions providing traditional methods of instruction.

In the past, there were a few correspondence courses and a few independent study opportunities, but by and large, the system did not significantly change for 200 years. Now, the widespread availability of the Internet and information search engines has changed access to information. Technology is infusing into the educational system a new competitor that will steadily reshape the delivery of higher education: distance learning. The University of Phoenix Online is now the nation's largest accredited university, with 17,000 instructors on 170 campuses. This virtual university has conferred over 171,000 degrees since 1976 and currently has nearly 300,000 students enrolled in degree programs. The company began offering online degree programs in 1989. In fewer than 30 years, the University of Phoenix has grown to become the largest university in the nation!

This is the kind of growth sought by capital deployment from the private sector. Phoenix has set sail with the winds of techonomics at their back. Distance learning appeals to the growing retraining needs of an aging workforce, provides 24/7/365 delivery to anyone, anywhere, eliminates labor costs except for faculty, and minimizes most costs associated with housing and teaching. This is virtual education, totally customized and presented on demand by the finest instructors in the land. The resulting approach is highly efficient, highly convenient, and highly extensible.

These growth numbers for professional training prompt the question, Will distance learning make inroads into the student population going directly to college from high school (or for that matter, for K to 12)? Online forums allow for intellectual exchange and collaboration, but virtual university students still forfeit the social opportunities offered by the traditional approach, as well as face-to-face interaction with professors and classmates. Even so, hundreds of thousands of students are choosing the virtual university as their source of higher education. Businesses like the University of Phoenix succeed because they find a need not being filled, and they step forward to fill it.

Today's need is continuing education leading to degrees for many who are already in the workforce and must deal with logistical constraints. Tomorrow's need will be economical delivery of specialized higher education for a new generation of students. Traditional, brick-and-mortar institutions will have to adapt. Many of them wish to add a virtual dimension to their teaching, but few currently wish to be entirely transformed into virtual universities. They want to keep their brick and mortar, their systems of tenure, and other privileges. They review and accredit each other with the announced goal of keeping educational standards high — but notice that they also wish to maintain the status quo of academic privilege.

Institutions of higher education are reviewed and accredited by organizations comprised of university faculty and administration. The review boards share with those they review an inherent desire to perpetuate the current brick-and-mortar based system. While working to maintain standards, they also propagate salary requirements, floor space requirements, curriculum standards, etc. that increase costs (tuition or taxpayer), constrain creativity, and protect arcane degree programs with limited societal value except to the department itself. This form of self-evaluation is a form of protectionism leading to perpetuation of methods past their prime. The mindset of protection, rather than performance, still permeates the system.

Consider the tradition of academic tenure (from Latin tenere, to hold: at most universities a tenured faculty member can hold her job, guaranteed, with little fear of losing it — unless she commits a felony). Although some universities have taken steps to prod complacent, tenured faculty into productivity, the practice of tenure is based on protectionism, and tends to weaken the climate of competition and rewards based on actual labor and achievement.

Resistance to changing the old system of education is problem enough, but consider again the theme we originally looked at in this chapter: the poor record of U.S. education in science and math during recent decades. By contrast, international trends in technical education are looking up. Whereas, in the year 2000, there were only 60,000 U.S. engineering graduates, there were in China alone 800,000 graduates.[13-14] Japan produced 200,000. If technology is the creator of value in the emerging digital age, what do these startling numbers tell us? Who is developing the deepest sources of mind power available to capture future economic opportunities?

It is not hard to see why companies are relocating manufacturing, design, software development, and support to the Far East. Not only do the economics of labor costs work in their favor, but also the sheer volume of available mental capital is beginning to dwarf the U.S. system that spawned foreign institutions of education. For 30 years, graduate technical programs in the U.S. were hosts to international students, many of whom stayed to teach and practice here. Those who returned home have established and grown programs in their own countries. These programs far exceed our own in numbers of graduates — and perhaps it is only a matter of time before they exceed us in quality, as well.

Field: Education, higher.
Missing economic natural selector: Evaluation and change actively penetrating from outside the walls of the academy.
How imposed: Collegial reinforcement of the traditional academic worldview results in a self-review process, and a hiring process, that seeks self confirmation and attachment to brick-and-mortar classrooms as the most effective teaching vehicle. The combined processes of tenure (individuals) and accreditation (organizations) serve as a filter for a single-minded worldview that resists change.

Anticipated techonomic result: Anticipate continued significant growth in distance learning and in the gap between the number of domestic and international graduates in technology fields.

Techonomic recommendation: Eliminate tenure as a foundation of the personnel system and infuse accreditation boards with private sector reviewers. (Tenure does not exist in the private sector with the exception of some collective labor agreements in failing industries.)

GOVERNMENT: TECHONOMIC EFFECT IN MACROECONOMICS

Once people realize they can vote themselves money — that will herald the end of the republic.

— Ben Franklin

A democratic government is the only one in which those who vote for a tax can escape the obligation to pay it.

— Alexis de Tocqueville

This book has focused on how technological innovation, in competitive economies, has influenced organizations at all levels throughout history. At the top of the organizational food chain is national government. All organizations — including businesses, nonprofit enterprises, civic organizations, franchises, cities, counties, states, and ethnic groups — ultimately agglomerate into a nation, the greater community. From the wide-angle perspective of the national view, two twenty-first-century techonomic trends appear to be at odds:

1. **Techonomics Third Law predicts increasing efficiency.** That means more productive output from less labor input. At a company level, this means more outsourcing to more suppliers. Growth occurs by franchise or network expansion. At a national level, this may also mean more international outsourcing. But the question remains, where will the nation's productive value creation be sustained?
2. **Growth is the techonomic measure of community.** If the national government is the community, then for it to be "healthy," it must grow. If government continues to grow when its supporting organizations are diminishing due to competitive productivity improvements (fewer people to accomplish more output to remain competitive globally), there will come a time when government growth cannot be sustained by the financial burdens it places on the productive organizations within its jurisdiction.
3. **Perception is reality: the marketer's creed.** The short-term answer to managing this conflict between increasing private sector efficiency and growing government size is to manipulate perception of the national

economy using twenty-first-century techonomic trends. As long as there is a perception of strong growth and improving productivity, the government and the governed will move forward together in a state of economic bliss. Adept economic application of the First Law (exponentially decreasing electronics cost) has provided the government with the means to constrain inflation (Consumer Price Index) while increasing productivity (Gross Domestic Product) — virtually by controlling the statistical perception. Two concepts, the chained dollar and the hedonic adjustment, have been quietly introduced into calculation of government statistics in order to manage global perception of our economy. While what follows is difficult to comprehend (intentionally made so by its inventors, I think), this information provides the framework of today's perceived economic progress.

The Consumer Price Index (CPI) monitors the nation's rate of inflation by tracking typical living expenses over time. When the CPI increases too rapidly (more than about 4% annually), the purchasing power of currency is diminishing rapidly, and the cost of borrowing money increases. When the CPI is very small, or negative (less than 0.5% annually), then the currency is deflating, and the economy is in recession. The CPI is closely watched as a leading indicator of money supply and strength of the economy. An ideal target for the CPI is between 0.5 and 2.0%, showing healthy and controlled money supply. Many private-sector labor contracts and benefits plans are automatically tied to the CPI. Likewise, many cost-of-living adjustments for large government programs like Social Security and welfare are linked to the CPI.

For years, the term "19XX inflation-adjusted dollars" was used when comparing economic statistics over time. The inflation-adjusted dollar was a value calculated annually based on the cost of buying the same goods and services each year, then recalibrating for the change in cost due to inflation. A simple CPI example might include the sum of buying 50 loaves of bread, 200 gallons of milk, 1000 gallons of gasoline, and 1 year's rent on a 2-bedroom apartment. Figure our simple CPI each year, and then take the percentage difference between 2 years of interest to find the inflation adjustment to apply to other economic statistics. For most of the twentieth century, inflation-adjusted dollars were the basis of the CPI.

In 1995 to 1996, a new method of calculating "inflation-corrected 19XX dollars" became the approach to comparative value trends: the chained dollar. Whereas the inflation-corrected dollar measured the actual price of specific quantities of goods, the chained dollar compares utility replacement costs — the replacement of the same capability annually. Though subtle, this modification to the calculation of CPI has far-reaching and significant implications due to techonomic trends. This new calculation allows the economic efficiencies of Moore's Law to aid in the containment of inflation by including a significant technology component. Here is how.

Buying a gallon of milk or gas in 1995 and 2005 is a straightforward economic comparison. Buying a computer in 1995 and 2005 is not a straightforward economic comparison. Due primarily to Moore's Law, the $2,500 desktop computer in 1995 had one megabyte of memory, a 100 megabyte storage disk, and a 66-megaHertz clock. By 2005, a desktop computer cost about $2000 and had 10 megabytes of memory, a 40-gigabyte storage disk, and a 1.0-gigahertz clock. The chained dollar includes this in the CPI by approximating what the 1995 computer would cost if available today. A personal computer with capabilities 10 years old would cost about 1/32 of the cost 10 years ago, according to Moore's Law. But the chained dollar calculation never considers that a ten-year-old computer system would not be able to run current software or interface with available peripherals, rendering it useless.

While the technology component of the CPI is only a small slice of about 6 to 8%, this comparison of utility rather than actual device accounts for a reduction in the CPI of about 2% a year, before the price of life's staples are included. Since the CPI seldom exceeds 6% (except in the worst inflationary periods), the built-in technology reducer of 1 to 2% makes a significant difference in the annual CPI, making it lower than arrived at by the old method of "inflation-adjusted dollars." The perception of reasonable inflation is maintained even though the basic necessities (food, fuel, housing) are rising in cost at a significantly greater rate than the composite CPI indicates.

The Hedonic Adjustment uses the same techonomic trend (Moore's Law) in a different way to bolster the perceived productivity of the nation: calculation of the Gross Domestic Product (GDP). The GDP measures the value of goods and services produced annually. GDP is a fundamental statistic used to determine productivity per capita, limits on budget expenditures, target ranges for cumulative national debt, and the comparative condition of a national economy. During the process of creating techonomic metrics for domestic markets over the past 50 years, I kept observing a data discontinuity in the 1995 to 1996 timeframe that I could not understand. The introduction of the hedonic adjustment was the basis of GDP calculations that resulted in abrupt changes in GDP trends.

According to About.com, hedonic means "of or relating to utility (literally, pleasure related). A hedonic econometric model is one where the independent variables are related to quality."[15] In layman's terms, the hedonic adjustment is an accounting note that prices *today's technology economic utility* in terms of performance levels per dollar expended in 1996. Ten years with a Moore's Law performance doubling every 2 years means a hedonic adjustment approaching 30 times. **This hedonic accounting adjustment, based on technology advance, boosts the GDP by billions of dollars — virtually.** It accounts for money never spent for productivity that may or may not ever be used.

David R. Kotok of Cumberland Advisors gives a specific example from the second quarter of 2003, a particularly good quarter for the GPD. He writes,

> A single component of business fixed investment accounted for more than its (the GDP's) overall increase. Investment in computers soared by $38.4B, or 12 per cent, from $319.1B to $357.5B. The trouble is that much of this boom-like increase in computer investment never occurred. The apparent surge is a consequence of the hedonic deflator that U.S. government statisticians use when measuring computer output and investment. The aim is to capture quality improvements by calculating how much it would have cost in 1996 to buy a computer of equivalent power to today's machines. Measured in current dollars, however, this spending rose a lackluster $6.3B,

from $76.3B in the previous quarter to $82.6B — far below previous peak levels. In other words, hedonic pricing produced $32.1B of GDP in real terms, about 43.9 per cent of the reported second-quarter GDP increase of $73.1B. In its absence, GDP would have grown a mere $41B, implying a growth rate of 1.68 per cent. The important thing about hedonic pricing is that it measures dollars that nobody pays and nobody receives.[16]

You may need to read Mr. Kotok's explanation of the hedonic adjustment a few times. Simplifying and summarizing: if a customer bought $1000 worth of computing systems in 1995 and also bought $1000 worth of computer systems in 2005, the impact on the GDP of the purchase in 2005 would be adjusted to approximately $30,000 because of the hedonic adjustment (the computer capability bought by the customer is 30 times greater than a decade ago). The hedonic adjustment accounts for capability and converts it into currency, all in the name of substantiating increasing productivity.

The bottom line of Mr. Kotok's analysis is that the adjustment is getting very large now, measuring economic transactions that never took place. Does the GDP matter? Yes. National budget limits, spending limits, debt limits, and international economic standing are tied to the GDP. If the GDP ceases to grow, so does the government's access to money, from debt limits, budget deficit limits, and international investment confidence. Virtual growth is better than no growth at all; just look at the perception of the European economy measured without the benefit of a hedonic adjustment. Given the exponential nature of Moore's Law, the further removed from the 1996 hedonic adjustment calibration, the more virtual impact this factor will have.

In the decades ahead, trends toward globalization of businesses hold another serious consideration for governments. Businesses are tied more to world markets than to national or regional loyalty. As companies trend toward virtual operations, their ability to move locations for production, distribution, and headquarters becomes easier, more flexible, and primarily based on competitive economic considerations. Where to locate business operations becomes a question of economics and logistics, not national sentiment. Raise taxes too much, and companies remove their headquarters or operations to more favorable locations. Raise labor rates and entitlements too high, and businesses move production operations to more favorable regions. Combine the global communications infrastructure (Second Law), the mental augmentation of skilled labor (First Law), and continuing productivity pressure (Third Law), and the government must adjust its approach. At least, it must do so if it is to remain relevant and if its constituencies are to remain competitive.

Field: Government.
Missing economic natural selector: The government is a monopoly with no perceived equal. It faces the difficult challenge of maintaining the standard of living for its populace without commensurate improvements in productivity. The U.S. perception of military invincibility is masking economic deterioration.
How imposed: The two-party political system has long served the U.S. with a frequent adjustment to the balance of power. The polarization of the

worldviews between the conservative right and the liberal left has promoted ideology ahead of national unity. A malleable and often uninformed moderate center sways elections based on the media prowess of the candidate. Incumbents perpetuate the status quo because a very high percentage of them are reelected.

Anticipated techonomic result: Government will grow until the supporting constituents — the voters, taxpayers and businesses — demand change or are unable to bear its burden. If history is any guide, the cycles of national birth, growth, sustenance and demise are a part of fabric of this world.

Techonomic recommendation: Government is faced with growing entitlements, a business tax base that can readily relocate operations and employment, an aging populace, a growing population of illegal immigrants, and terrorist threats that have been destructive domestically and in many countries around the world. Techonomic recommendations for government include:

1. Use a transaction tax system (Third Law). A tax system based on transactions, commonly known as a consumption tax, will be easier to collect and administer in the virtual world. A consumption tax will not be a catalyst for companies to relocate operations internationally as, increasingly, business taxes are. And even if businesses do relocate, tax will be collected on their domestic transactions.

2. **Enforce the law.** For example, the debate over illegal immigration policy overlooks the simple fact that the nation is allowing illegal practices. Whether the national economy benefits, humanitarian respect requires, or states rights are overrun, the point is that nonenforcement of immigration laws is clear for all to see and sends a very bad societal message. National policies, whatever we make them, must be codified in law and enforced if a fundamental respect and trust for government is to be maintained with the populace. A generation of Americans is clearly observing that many key laws have no meaning. This is a path to chaos.

3. **Term limits.** Government office must not become a career path, but an avenue of service born from individuals of diverse training and experience. We cannot have a broad perspective to address and solve national issues when the professional background of the overwhelming majority of politicians is only from one field — law.

4. **Encourage personal productivity.** The high standard of living in the U.S. has spawned a level of entitlement that is not sustainable. Even the safety net must be earned with some form of service to others within our society, not unlike the Civilian Conservation Corps established during the Great Depression. The many projects of the CCC stand today in our national parks as monuments to a people who cared for their citizens and developed a responsible work ethic while building the nation's infrastructure.

SUMMARY

COMPETITION IS THE TECHONOMIC EQUIVALENT OF NATURAL SELECTION

In Darwin's Theory of Organic Evolution, natural selection was the filter that caused organisms with "environmentally preferred" traits to survive and pass on those traits to the next generation — survival of the fittest. In the Techonomic Theory of Organizational Evolution, market competition is the filter favoring viable technologies and the organizations that intelligently employ those technologies and best practices.

ORGANIZATIONAL EVOLUTION FAILS WHEN COMPETITION IS REMOVED

Without market competition to spur best practices, an organization becomes ineffective relative to other organizations that operate in a competitive environment. Competition is removed by a number of means, including monopoly, overregulation, protectionism, and lack of economic accountability.

THREE DOMESTIC MARKETS DEFY TECHONOMIC TRENDS

Three key domestic markets — energy, healthcare, and education — have been shielded from competitive pressure and are exhibiting declining productivity trends.

ENERGY PRODUCTION ECONOMICALLY INHIBITED BY DOMESTIC REGULATIONS

Environmental policy has made domestic energy production economically noncompetitive with international sources; hence the massive and increasing energy trade deficit and energy vulnerability for the U.S. Among many possibilities, government action could include opening up domestic areas for production and refining of domestic energy sources; capping liabilities on nuclear production facilities so that these facilities can be capitalized, licensed, built, and operated domestically; and funding rapid development to commercialization of inexpensive solar-voltaic energy devices.

HEALTHCARE SYSTEM LACKS PERSONAL RESPONSIBILITY

Every player in the healthcare community — including insurance providers, lawyers, hospitals, and healthcare professionals — is attacked as the root of the problem. But the real problem is the lack of personal responsibility for the economic impact of healthcare choices. The system must return to personal payment for insurance and care, and to open-market pricing of services. Under the current system, someone else pays the insurance, pays the healthcare bills over a minimal deductable, and ultimately makes the healthcare decisions. This is flawed to the core and is becoming a significant economic burden, compromising the rest of the U.S. economy.

Education System Perpetuates a Failing Monopoly

K to 12 education student performance is now on par with the third world. The monopoly is perpetuated by an alliance between government and labor that refuses to accept competition as the only foundation for lasting, positive change. The system is also overburdened with social agendas far beyond the original mission of education. Vouchers offer a means to introduce healthy competition into the K to 12 system. These would revitalize the system by opening it to competition from a wealth of neighborhood, community, and religious sources.

Government Accounting Adjustments Mask Performance

Since 1996, the chained dollar (CPI adjustment) and the hedonic adjustment (GDP adjustment) have used economic trends predicted by Moore's Law to mask the actual performance of the economy. Many broader economic actions are linked to these annual values, so the implications of these adjustments are far reaching.

Key Terms

capital	capital formation
free market	competition
Consumer Price Index	inflation-adjusted dollar
chained dollar	Cost of Living Adjustment
Gross Domestic Product	hedonic correction
best practices	continuous improvement

QUESTIONS

1. Competition is the key to progress. Apparently, the government agrees on some level, because it tries to assure competition by enforcing antitrust laws and breaking up monopolies and collusion in closely knit industries. In one of the most notable antitrust cases, American Telephone and Telegraph was separated into the "Baby Bells" just over 20 years ago (1984). Consider the rate of change in communications prior to and since the "breakup," and draw conclusions on the importance of competition. (For starters, Google: AT&T breakup.)
2. In 1989, with the fall of the Berlin Wall, an entire "superpower" culture pulled aside its iron curtain, exposing a superpower in military appearance only. The socialized, centralized economy was in shambles. Survey the status of Russia today; make and rationalize any observations regarding the lasting effects on society of a prolonged economic system that is devoid of competition. Do you have any observations about the acceptance of (reintroduced) capitalism in the former Soviet-block countries?
3. A host of alternative energy sources are being considered for the future, including wind, wave, solar, fusion, hydrogen, and coal gasification. Do a techonomic analysis of one of these energy sources. Determine the

current economic cost of producing a kilowatt-hour (kWh) of electricity from this source and the type and magnitude of technology advance needed to make your energy source economically viable. Are the advances aided by the Laws of Techonomics, or are they totally unrelated?
4. By some estimates, the healthcare sector now exceeds 15% of the Gross Domestic Product. The average annual insurance premium for a family of four is approaching $10,000, which also happens to be about the minimum wage for a 2,000-hour work year. At what point would you, a business owner, opt out of the insurance program and simply increase employee salaries while expecting them to take care of their own healthcare plan? With wages increasing at 2 to 4% annually and healthcare insurance premiums increasing 10 to 15% a year, there will come a point when every business owner will be compelled to make this choice!
5. Grading on the ACT and SAT standardized tests has been adjusted in the last decade to improve the perception of test scores (and reduce scrutiny on the failing performance of the education system). International standardized test results reveal the poor performance of U.S. students in math compared to the rest of the world. Short of vouchers, what approach would you recommend to restore the educational system as a whole? Where is the incentive within the current educational business structure to perform and improve?

REFERENCES

1. U.S. Census Bureau, Trade in Goods (Imports, Exports and Trade Balance) with China, Foreign Trade Statistics, U.S. Census Bureau, 2005, http://www.census.gov/foreign-trade/balance/c5700.html.
2. U.S. Department of Labor, Summary of Labor Statistics, Summary 00-16 August 2000, Bureau of Labor Statistics, U.S. Department of Labor, 2005, http://www.bls.gov.
3. Moody, E.F., Jr., Life expectancy tables, 1998, http://www.efmoody.com/estate/life-expectancy.html.
4. Kaiser Family Foundation, Trends and Indicators in the changing health care marketplace, Health Care Marketplace Project, Henry J. Kaiser Family Foundation, 2005, http://www.kff.org/insurance/7031/print-sec1.cfm.
5. Centers for Medicare and Medicaid Services, Historical National Health Expenditures Aggregate, Health Accounts, Centers for Medicare and Medicaid Services, U.S. Department of Health and Human Services, 2003, http:// www.cms.hhs.gov/NationalHealthExpendData/downloads/tables.pdf.
6. Toosii, Mitra, Office of Occupational Statistics and Employment Projections, Bureau of Labor Statistics. *Consumer spending: an engine for U.S. job growth*, pg 3, Table 1: Personal consumption expenditures, 1980, 1990, and 2000 (actual) and 2010 (projected), http://www.bls.gov/opub/mlr/2002/11/art2full.pdf.
7. Kaiser Family Foundation, Premiums increased at five times the rate of growth in workers' earnings and inflation, Henry J. Kaiser Family Foundation, 2004, http://www.kff.org/insurance/chcm090904nr.cfm.

8. Kaiser Family Foundation, *Concentration of health spending in the total U.S. and family populations,* Health Care Marketplace Project, *Henry J. Kaiser Family Foundation, exhibit 1.11, 2002,* http://www.kff.org/insurance/7031/trends_updates.cfm.
9. Hodges, M., U.S. education productivity index, Grandfather Education Report, Grandfather Economic Report, 2005, http://mwhodges.home.att.net/education.htm.
10. OECD, Programme for International Student Assessment, Organisation for Economic Co-operation and Development, 2003, http://www.pisa.oecd.org/dataoecd/0/48/33995376.xls.
11. Young, B.A., Public school student, staff, and graduate counts, school year 2001–02, National Center for Educational Statistics, Institute of Educations Sciences, U.S. Department of Education, 2003, http://nces.ed.gov/.
12. Hill, N. and Cornwell, R., *Think and Grow Rich: The Original Version, Restored and Revised,* Aventine Press, San Diego, 2004, 81.
13. Yoo, I.-S., National Society of Professional Engineers, USA Today, 4/20/04.
14. China's cheap, high-quality labor lures foreign investment, China Internet Information Center, tom.com, http://us.tom.com/english/3297.htm.
15. Moffatt, M., "Hedonic," Economics, About, About Inc., 2005, http://economics.about.com/library/glossary/bldef-hedonic.htm.
16. Kotok, D., *Update re: Stocks, Bonds, and Cash, Cumberland Advisors, 2006,* http://www.cumber.com/commentary.aspx?file=091303.asp&n=1_mc.

9 The Techonomic Future

You can observe a lot by just watching.

— Yogi Berra

EXPANDING THE BOUNDARIES

As we have seen, techonomics recognizes early-stage military development as a frequent indicator of technologies on their way to broader markets. Military research often expands the boundaries of science while exploring new physical principles for military use. Digital computers, software languages, integrated circuit electronics, atomic power, satellite communications, global positioning systems, radio-frequency identification, medical isotopes, lasers, and the Internet were all developed, or significantly advanced, for military applications prior to commercial popularization. As we have also seen, the reason this happens now and has happened throughout history is simple: fear is a great motivator; the desire for safety is paramount. The military in times of war becomes a focus of resources, both financial and intellectual, without rival. Military innovations are then adapted for applications in the larger commercial marketplace.

Military applications cover many dimensions: altitude (jets, satellites), depth (submarines), speed (communications, ballistics), computation, human cognition, training, remote controlled operation, virtual reality, and power — to name a few.

Military technologies on the horizon now allow us to glimpse possibilities that lie ahead in the next 20 years. We must remember the trends of the first three laws of techonomics, particularly electronic and communicative ubiquity, as we ponder the commercial possibilities of these technologies for the future.

THE EXPANDING THIRD DIMENSION

Multidimensional extension of current systems offers new performance horizons for technology. The two-dimensional plane becomes the three-dimensional volume. The large becomes small, and the small becomes minute in space or time. The hot gets hotter beyond previous ranges, and the coldest cold unleashes amazing material properties. As we search these trends in technology, a new horizon of emerging possibilities comes into view. When these paradigm-shifting technologies find their place in the mass market, the result is tremendous economic value and changed standards of living.

Humans commonly work in the two planar dimensions, length and width. For the most part, we grow crops on one layer of precious two-dimensional land. We watch television and movies on a two-dimensional display. We write and print on sheets of two-dimensional paper. We play chess on a two-dimensional chessboard.

Now that we have the mental tool of the computer for design, control, and operation, three-dimensional implementation of many technologies is emerging. Following are a few three-dimensional directions with future consumer potentials.

Computational Cube or New Platform?

Since their introduction in the 1960s, most integrated circuits (ICs) have been designed and manufactured in a planar geometry. Typically, two-dimensional ICs were then placed on two-dimensional boards and packaged into three-dimensional electronic devices. This approach was the standard practice because the third dimension created design, manufacturing, and heat-transfer challenges. As long as Moore's Law could advance the capacity of silicon flat planes every 2 years, there was little need to seek alternatives.

But IC design and manufacture has now progressed to the point where physics is beginning to limit how small traces and elements can be. Also, formidable quantities of capital are required for ever-more-precise and clean fabrication facilities. Because these increasing costs are returning diminishing performance, we can anticipate greater effort in exploiting the third dimension (depth, combining multiple layers of single-plain silicon into a high-performance unit). Also, optical or organic computation schemes will emerge to compete with silicon as the basis of electronic computation.

Vertical Growing and Living

The long view of history reveals technology enhancement and population growth walking hand in hand. Without power to transport crops to urban areas, the most populous cities would never have exceeded a half-million people. Without mass use of automobiles and electricity, centers of population could not have continued growing. Did technology advance cause the growth, or did the growth demand new technology solutions? Then as now, the technologies that would solve the problems were emerging before the general population demanded a solution.

Our key cities have grown vertically due to rising land cost and migration patterns that have brought many to urban centers in search of opportunity. Increasing population density places a greater challenge on crop production, which also competes for suitable land. The three-dimensional answer is vertical, environmentally controlled agriculture. The potential advantages include localized growing of crops (minimizing transportation costs and maximizing freshness), a year-round growing season, reuse possibility for existing unused buildings, and increased food production density per acre.

Economic success for this activity depends on the energy efficiency of the approach and the quality/quantity of agricultural output afforded by three-dimensional agriculture. Techonomic keys to success include:

- **Efficient solar spectrum lighting**. Required to produce the full-spectrum light necessary for growth and ripening of fruits and vegetables. It must be efficient to be cost effective from an energy standpoint. Facilities using

high-temperature inert gas lights, combined with optical light pipes to distribute light throughout a facility, are in the experimental stage.
- **Vertical planting and care methods**. Needed to feed, water, grow, and harvest crops in a three-dimensional manner. Such methods will succeed only if labor costs for production are contained within current two-dimensional production standards. Otherwise, cost will not be competitive and the method will not see widespread acceptance.
- **Genetically engineered plants**. Needed to optimize for growth and quality within a controlled environment. Research, development, and production of genetically engineered plants will expand rapidly if vertical agriculture starts to flourish.

The governing factor in three-dimensional agriculture will be economics. The technological hurdles are small, but the economic questions loom large: facilities costs, conversion costs, energy costs, labor costs, and crop productivity relative to traditional methods. Advancing technology may reduce many of these costs while the cost of land use in traditional agriculture rises. Perhaps there is a future techonomic crossing point.

Three-Dimensional Entertainment

Entertainment based on technology has been "flat" (two-dimensional) for a long time. Books, newspapers, magazines, paintings, and more recently TV and cinema, have all been presented on essentially flat surfaces. Today, three-dimensional and immersive media are emerging in laboratories worldwide. Some of these developments are more promising than others, but all are "eye opening":

- **Holographic projections** are three-dimensional images that can be projected into a volume and observed from any direction. Advances in computer-generated models and holographic projection, color representation, and contrast are progressing to make these images increasingly realistic.[1] Development of real-time 3D motion capture and display may transform the future of television.
- **Stereo projection and video displays** are emerging to provide a realistic depth perception of motion video. Used for remote operations and specialized group entertainment today (Disney World), stereo viewing systems rely on timed shuttering, polarization, or color filters to separate images, providing parallax from two cameras. The cost, complexity, and ergonomics (human individual perception differences) of these systems have limited their adoption. Beyond the wizardry, they serve little tangible benefit in mass applications and are not expected to find mass appeal.
- **Near-eye display systems**, also referred to as head-mounted displays, are nearing the quality and cost point needed to impact the consumer marketplace. These devices incorporate very small displays with focusing optics providing high-resolution images. The devices are presented in a configuration like eyeglasses. Several technologies — including miniature

light-emitting diode (LED) displays, color-wheel light-collecting diode displays, and direct light-scanning-to-retina displays — are competing for market viability. Organic light-emitting diodes (OLEDs) are nearing the marketplace for these small displays. These screens are composed of a layer organic material that changes color in the presence of charge. Various configurations and applications include: monocular (one-eye) displays offering a compact portable viewing system for Web sites and instruction manuals; semitransparent displays allowing the overlay of computer-generated information coordinated with location in the real-world environment; and totally immersive systems that occlude the ambient world and transport the user into a virtual world (for video gaming, instruction, etc.) or another location (telecommuting). Technology issues of resolution, power consumption, and wide field of view are being addressed as the economics are improving with expanded use in military applications like night vision sights and commercial camcorder eyepieces.

- **Computer graphic representation** for three-dimensional displays has advanced with the computational power of available computers. The graphic power of a $250 video game platform today exceeds the real-time video rendering capability of any computer available 20 years ago. The first mass applications have been experienced in commercial movies, where the separation between real and virtual is increasingly difficult to perceive. Progress in realism of video game representations has also advanced significantly. The move from three-dimensional rendering to a two-dimensional display will be followed by three-dimensional rendering to a three-dimensional display once the display technologies are commonly available. The military and key industrial uses have already begun, and the mass application may result from a "killer application" in the video game market — networked first-person-adventure games.

Three-dimensional entertainment will be a major transition, like the conversion from black-and-white to color television. Three-dimensional entertainment will require a new infrastructure for broadcast, a multitude of new electronic sets, and scores of new programs, just as color television needed in the 1960s. A generation later, the technological advances of cable and satellite signal distribution led to two more orders of magnitude of available channels for new specialty programs. One more generation later, digital distribution over broadband, and new technological products for real-time capture and replay of programming led to mass customization of individual programming selection.

Simultaneous with these television advances, the video game market was expanding as the underlying technologies provided an ever-more-compelling experience. From the teletype-printed Xs and Os of the early *Star Trek* games played on a monolithic centralized computer to the real-time, three-dimensional, color, interactive videogames like *Halo,* the videogame experience has advanced to the point of addiction for some participants. Networked games now allow entire virtual communities (of thousands) to assume pseudo-identities and compete regularly with people they have never met face to face, playing compelling games of endless duration such as *Everquest*. So compelling is today's video game experience that academic journals, conferences, and degree programs for video gaming have sprung up. Gaming can be used for instruction and intellectual development as well as for fantasy.

Ergonomic display systems will provide a total sensory immersion: stereo viewing and stereo audio coordinated and directed by the user's head orientation. *It will not be long before more than 95% of sensory data coming to a person's senses from the external world can be emulated by a computer simulation.* The participant will temporarily become part of a virtual, computer-generated world, interacting remotely with others within an environment that includes sight, sound, motion, and speech. For education, entertainment, communications, and manipulation of the masses, this convergence will far surpass any preceding media experience.

Three-Dimensional Printing

Although the field of three-dimensional printing is in its infancy, it is a promising-looking infant. "Three-dimensional printing" refers to any of a number of techniques used to create a three-dimensional tangible object from a three-dimensional computer model, without human intervention. The concept is simple. Imagine an ink-jet printer that uses molten plastic rather than ink. Spraying numerous successive layers of the plastic onto a two-dimensional surface, the printer (under computer control) can generate a three-dimensional object. Two types of plastic are used: one stays and one leaves voids by shrinking or getting washed away (water soluble) after the process. This technology was developed for rapid prototyping of parts to reduce the time and cost of going from concept to tangible part. Emerging systems can use zinc alloys as a material option to "print" a real part, strong enough for use in a product. Older methods of three-dimensional printing called stereo lithography were complicated and expensive and produced flimsy parts. The new methods produce parts simply, quickly and relatively inexpensively — techonomics in action!

Need a new doorknob? Print a doorknob. Need a toy racecar? Print a toy racecar. Need a fork, spoon, plate? Print them. Single-material, three-dimensional printers are already making tools with working gears and bearings with working roller balls.

Future advances in this approach may lead to on-the-spot-manufacturing for military purposes. Imagine a three-dimensional printer combining plastic, metal, gel, insulation, etc. It will "print" a complex object, including its internal electronics, from a project file describing the object. This is another area of technology that will develop from military applications. This particular technology will leverage from price reductions *a la* Moore's Law, just as two-dimensional, jet-ink printers have significantly reduced in price over the last two decades. Further information on three-dimensional printing products can be obtained from Stratasys, Inc. or by searching the Web for "three-dimensional printing" or "rapid prototyping."

GETTING REALLY SMALL

Human imagination — and sweat — extend the boundaries of space and time every day. One direction of great interest to scientists is the small — the very small. The growing field of nanotechnology (where things are measured in units of 10^{-9} meters) studies the very small in terms of space. Scientists looking at small in terms of time measure things by the femptosecond (10^{-15} second). These measurements of distance and time are so tiny that the layperson may have difficulty comprehending them. Nevertheless, studies in smallness represent the next frontier where mankind will produce more utility with less physical material.

Ironically, instruments must get very large to study the very small. The Spallation Neutron Source (SNS), currently under construction at the Oak Ridge National Laboratory, is a gigantic particle accelerator that will be used to unlock greater secrets of material properties on the smallest scales. The $1.4 billion project is scheduled for completion in 2006 and will serve the world's scientific community with an instrument that stretches for almost half a mile to unlock material secrets in the physical realm of the very small.

Space: Nanotechnology

A nanometer is one billionth of a meter. The diameter of a human hair is on the order of 20,000 nanometers. The diameter of a hydrogen atom is about 0.1 nanometers. The field of nanotechnology is engaged in creating and understanding the world in the range of 1 to 1000 nanometers. Nanotechnology is a field in which scientists are creating objects by the specific and individual arrangement of atoms or molecules. According to Dr. Ralph Merkle, nanotechnology in the form of "molecular manufacturing" pursues the following:[2]

1. Getting essentially every atom in the right place.
2. Making almost any structure consistent with the laws of physics that we can specify in molecular detail.
3. Keeping manufacturing costs from exceeding (very much) the cost of the required raw materials and energy.

Molecular embossing has already been demonstrated by International Business Machines; they have been able to emboss their logo one molecule at a time on a flat plate. As the technology advances, scientists envision "nanomotors" (extremely small electric motors), "nanogears," "nanopropellers," etc. with which to make complex and useful nanomachines. Think of the possibilities. For example, nanorobots could navigate the human body to perform surgery or deliver drugs directly to the internal organs without an incision. Actually, medical use of small machines is a present reality. For example, a camera/transmitter the size of a pill can now be swallowed and provide a picture of the gastrointestinal track during its journey through the body.[3]

But the *really* small is coming. Nanotubes (single-atomic-thickness-walled tubes) are exhibiting properties that resemble superconductivity (electrical conduction without resistance) and strength per unit weight that exceeds known properties of the strongest materials. Nanogels (spheres of sand with nanometer voids) provide significant thermal insulation in a material that is transparent. Nanocircuitry may hold the key to the continuing advance of Moore's Law after the limits of lithography are reached (now nearing 1-micron, or 1000-nanometer, feature size).

Time: Actions in Femtoseconds

In a femtosecond, the light will travel only about 3×10^{-6} meters. That is not very far. Today's fastest computers have clock cycles measured in one billionth of a second. That stretch of time consists of one million femtoseconds.

So why mention a femtosecond if it is a million times shorter than the timing cycle of today's fastest computers? Femtosecond operations are in the future of computers, weapons, and machining methods. By creating a laser pulse of 50 to 1000 femtoseconds, a directed energy beam can be used to remove a single atom of material without heating or vibrating the material near it.[4] Hence, femtosecond control of energy becomes a key means of nanomachining. Very short time leads to very small dimensional cuts. These machining techniques can be applied to small devices or to human surgery, providing a precision heretofore impossible. Pulsed energy weapons may become a key to crowd control, since pulsed energy could cause extreme pain for a short time without leaving permanent injury. Similar technology could be used to disable communications or computational equipment by creating much larger, but very brief electromagnetic interference.

Femtosecond: a word to remember. It is an extension of the time dimension available for future discovery. Track the clock frequency of personal computers with each successive generation, and you will maintain a perspective on humankind's grasp of the minuteness of time.

EXTREMES OF TEMPERATURE: ENERGY HOT AND COLD

Material and physical properties of matter change as they approach temperature extremes. The realm of fusion and high-temperature plasma reside in the very hot region, and superconductivity resides in the very cold (although the superconducting transition temperature for new materials is getting warmer with the march of materials technology).

Plasmas exist in many ranges of temperature and material density.[5] Plasma is the most common form of matter; over 99% of the known universe is plasma. Plasmas exhibit magnetic properties and are electrically conductive. These properties are continuously being explored and exploited for useful purposes. Applications on the market now, or emerging for the future, include:

- **Plasma lighting.** Emulates the sun's spectrum; may be important to future indoor agriculture.
- **Plasma sterilization.** Uses plasma to kill viruses and bacteria.
- **Plasma sheathing**. Allows plasma to dynamically cover a solid surface and change its physical properties (friction or surface reflection). One interesting characteristic of plasma sheathing is its ability to increase the sound barrier, allowing subsonic flight speeds to increase.

- **Plasma material treatments.** Uses plasmas to process the surface of materials permanently, changing their properties.

High-temperature plasma will no doubt have vast industrial application, but our current understanding of those applications is at an early stage, something like our early understanding of coherent light (lasers) in the 1960s. Consider the many and vital applications of laser technology today! Likewise, there will be many uses for plasma technology, but significant application progress is needed to determine cost effectiveness. Over the years, lasers progressed to everyday applications (printers, CD players, tools of all kinds, communication devices, etc.). Plasma technologies will find equally valuable applications to energy, environmental restoration, and material preparation in the years ahead.

On the other end of the temperature scale is research into superconductivity. Materials exhibiting superconductivity transfer electric current without loss (no resistance). Historically, certain materials have exhibited this property at extremely low temperatures (approaching absolute zero: 0K, or –273°C). Material breakthroughs over the last 60 years have led to a class of ceramic superconductors exhibiting this property at a crossover temperature of 138K, or –135°C.[6] The crossover temperature is important, because it determines the temperature below which the material must be maintained in order to exhibit superconductivity. The closer this temperature is to room temperature, the less system cooling is required to maintain superconductivity. One obvious goal of this research is to find materials that will superconduct at room temperature (~293K or 20°C), eliminating the need for system refrigeration, thereby simplifying superconducting devices while making them more cost effective. Since superconductivity eliminates heat loss from electrical resistance, the physical size of traditional electric devices can be shrunk significantly when implemented with superconducting materials. The future will reveal several worthwhile applications including:

- **Superconducting electrical transmission lines.** Inner cities have constrained spaces for utility infrastructure support, but these same areas suffer from an aging electrical infrastructure and increasing power demands. Superconducting transmission will pass more power through the same constrained space with less energy loss. Long-distance transmission from remote power generation plants to populated areas will also deliver more power with less loss.
- **Superconducting motors and generators.** These devices will function just like their traditional counterparts, with the important improvement that they will not generate heat from electrical resistance. This will allow the superconducting versions to be much smaller and more efficient.
- **Superconducting magnets.** Superconducting magnets can generate significantly higher magnetic fields in smaller volumes than traditional magnets due to their ability to conduct higher currents. Resulting applications include magnetic resonance devices for DNA research and drug discovery, beam-focusing magnets for microwave weapons, and levitation devices for vehicular travel.

- **Superconducting energy-storage devices.** Superconducting coils that store electric current until it is needed could replace many of today's batteries. Like a mechanical system storing kinetic energy in a flywheel, a superconducting coil would store current in the coil until the current was transferred to an external circuit. With no energy losses, and the ability to charge and discharge instantaneously any number of times, such a system would be ideal for something like regenerative braking on automobiles — if a room-temperature superconductor were to be discovered.

Unlike the progress in semiconductors governed by Moore's Law, where progress has been steady and predictable, the advance of technology related to superconductivity has been far from consistent or predictable. Material breakthroughs come by trial and error — and occasionally, by accidental good fortune. Scientific breakthroughs are much harder to predict than trends in engineering advances, making it difficult to anticipate the timeframe for something like room-temperature superconductivity. When room-temperature superconductivity in an affordable material is discovered, the proliferation of new devices to the marketplace will happen rapidly. There is a great deal of information about superconductivity available online.[7]

BIOLOGICAL PROCESSES

Human understanding of biochemistry is expanding as rapidly as our understanding of physics. The decoding of the human genome, cloning, metabolic imaging, and genetic engineering for plants and animals are all fields whose secrets are surfacing. Advancing computational power has been at the foundation of all these developments. Advanced computing enables scientists to mathematically analyze billions of combinations of DNA to discover key patterns. Advanced electronics magnify images to new levels of clarity, allowing for manipulative procedures that were previously inconceivable. High-sensitivity radiation detectors capture metabolic activity in three dimensions, allowing computers to generate models of internal functions and revealing, without a single incision, brain activities or tumor growth. Techonomics anticipates these advances will lead to several exciting applications, including:

- **Mass-customization of drug treatments for disease.** Combining DNA testing and drug development with metabolic imaging, the future of medicine will be customized, genetic pharmaceuticals. Drugs will be delivered straight to the source of illness with personalized DNA carriers that will adhere only to the diseased tissue. Treatments will be monitored by noninvasive imaging techniques, allowing doctors to make any needed adjustments without destroying the surrounding healthy tissue.
- **Predictive medicine.** Unlocking the full detail of the human DNA sequence will lead to accurate prediction and compensation for genetic diseases. Rather than waiting for symptoms of a genetic disease to emerge, treatments for many genetic problems may be solved *in vitro* or through genetic selection of egg and sperm.

- **Personal telemedicine**. As reliable, versatile, and inexpensive test methods for blood and other fluids are developed and interfaced to computers, routine healthcare access and cost could largely be circumvented via a personal, computer-based medic. Today, many people use the Internet for deeper understanding of diagnosis and treatments suggested by their doctors. Tomorrow, personal telemedicine will take us closer to autonomy. We will be able to touch a device that obtains vital signs and needed fluids and then makes a diagnosis, including prescribed treatment. The keys are automating a series of known tests and the associated data acquisition so that an expert system can reach a proper diagnosis and treatment for the patient — and then e-mail a prescription to the nearest pharmacy for delivery to your home within a couple hours! Medical practices are already using global services to read and interpret medical imaging, from X-rays to cytometry.
- **Replacement organs**. Will we be able to grow organs to harvest for our health needs? The moral implications will have to be addressed alongside of technology challenges. Such a possibility is conceivable, and more than likely achievable.

VANISHING INTO THE VIRTUAL

We have briefly looked at boundaries where scientific advances will result in future generations of techonomic opportunity. Dimensional ranges of space, time, and temperature will expand to unlock more physical secrets of the universe, leading to many technology developments, some of which will proceed to mass markets by withstanding the test of competition. The most important word to keep in mind as you analyze the opportunities of the emerging world is the word *virtual*.

In the virtual world, measured dimensions blur into the past. Your presence can be anywhere and everywhere at the same time. You have no mass, no temperature, no girth, and consume nothing while you make your presence known on demand anywhere you are welcomed. The twenty-first-century techonomic trends will combine to make your presence and your remote experience better, faster, more realistic, less expensive, and more compelling as the years go forward. Special effects in movies have already blurred, using computer graphics and "trick photography," the line between real and imaginary as presented on flat screens. The generation ahead will experience a range of life in the virtual world that rivals the life-experience in the real one.

Free of dimensional boundaries, we shall soon face possibilities that are actually or, rather, *virtually* endless. A range of virtual services are already 24/7/365, and you will have access to real or virtual representations of experts in every endeavor. Virtual education will provide virtual experiences of subject matter or virtual field trips to the places being studied. Virtual healthcare will use PC peripherals to monitor your condition and intelligent agents to route your case to a leading expert for your treatment (even if the expert on the other side of the screen is not live, the representation will be so compelling and comforting, you will not realize it!). Virtual entertainment will allow you to interact with hundreds of people you have never met, playing imaginary games you cannot play in the "real world," earning credits you can exchange for real goods, performing tasks that are taboo in the real world (already happening, but the future will bring a totally transparent and immersive interface, making the virtual very compelling).

Virtual evangelism will expand, allowing spiritual messages to be taken to all people of the world, reaching beyond territorial threats of violence. Virtual vicarious living will allow you to determine any role you want to play, and the virtual world will interact with your desires. The Virtual Age will lead to personal enjoyments and comforts beyond our wildest imaginations. It may also lead to personal isolation and self indulgences of all kinds. Personal isolation is the down side of the virtual juggernaut, but it is not slowing down the transition from real to virtual.

There are hundreds of business opportunities to be found in transferring a business implementation from the real to the virtual. Here are just a few contemporary examples of virtualization rapidly changing the world:

- Letters to e-mail
- Localized flea markets to global Web auctions (eBay, others)
- Physical storefronts to Internet storefronts (Amazon.com is the world's largest bookstore.)
- Paper and metal money to electronic transactions
- Passive barcode labels to active, IP-addressable RFID tags on products
- Live telephone operators and service technicians to automated conversant telephone systems
- Live insurance agents to electronic sale of insurance and remote processing of claims
- Real classrooms to interactive distance learning
- Live, human actors to computer animated/automated characters (already in video games)
- Personal social contact to virtual courtship
- Real hide-and-seek games to networked virtual adventure games
- Physical libraries to electronic books on demand
- Paper maps in the glove compartment to Global Positioning Systems (GPS) in the dashboard
- Physical competition for tangible prizes to virtual competition for virtual tokens that can be sold for hard currency on electronic auction sites

The virtual age is not a world of virtual reality but a world in which reality is virtual. In our glimpse of the future, separating the "real" from the "virtual" becomes an exercise in semantics. Is a store a place you physically visit to physically buy things with physical money? Or is a store a virtual website storefront that you virtually visit to make a virtual transaction with virtual currency — receiving something physical only when it is delivered at your home? Your answer matters very much. The advent of the virtual world opens opportunities to rewrite the rules of transactional engagement — to recalibrate the "playing field" for all commerce.

Parallels between the advent of the virtual world and the advent of the microprocessor are instructive. In the mid-1970s, the first microprocessors for commercial applications became widely available. At that juncture, even without full validation of Moore's Law, it was evident that these tiny, smart devices would eliminate many widespread and commonplace mechanical systems. On the automobile alone, mechanical speed governors, mechanical chokes, mechanical/hydraulic linkages to steering, transmission, and brakes, mechanical windows, mechanical locks, mechanical wiper controls, etc. have all been eliminated or simplified by microprocessors and electrical sensors/actuators. The same trends are true in devices and systems ranging from washing machines to record players. Each new application of a microprocessor to eliminate a mechanical system resulted in opportunities for new features, better performance, and lower costs. The trend toward "virtualization" offers similar across-the-board opportunities to rethink and reinvent industrial and business organizations and the societies they constitute.

Our guides the first three laws of techonomics substantiate the rapid approach of the virtual. Reconsider the virtual impact of these laws: (1) The Law of Exponentially Diminishing Electronic Cost (Moore's Law), (2) The Law of Exponentially Increasing Connectivity (Metcalf's Law), and (3) The Law of Increasing Productivity (Coase-Downes-Mui's Law). An exponentially increasing number of people have access to an exponentially increasing number of offerings at an exponentially diminishing cost in a continuously improving manner. Real-world offerings become the direct targets of virtual-world strategies in the same manner that mechanical implementations were the direct targets of electronic innovation. In the end, techonomics anticipates that the most successful organization in the marketplace is the one using the most effective technologies to economically deliver their endeavors.

Virtualization — moving business endeavors from the real to the virtual world, seeking the most cost-effective total process — is occurring on many fronts. The journey from real to virtual retail stores is just one example; consider a few more:

- **Virtual company, the micro-multinational.** Silicon Valley has been the bellwether of the U.S. entrepreneurial community for the last 40 years. A new phenomenon in Silicon Valley is the rise of the "micro-multinational," the virtual company. A small, core team developing the company concept characterizes this organizational structure. This team may obtain intellectual property protection, develop a prototype, perform manufacturing, establish distribution, and sell product — entirely via global outsourcing, seeking the most cost- and time-effective means to market. Hence the name micro (small startup) multinational (support services globally acquired). Today's labor cost differential between the U.S. and the Far East is large enough to justify the increased logistical challenge *even for the smallest of companies.* The hidden economic impact is that even if the micro-multinational is successful in growing beyond the startup phase, most of a venture capital investment and jobs are going to flow to international sources in order to support labor functions distributed around the world. No longer will U.S. entrepreneurial success convert to growth in the U.S. workforce, although it may convert to growth in the U.S. capital markets.
- **Virtual country, the macro-multinational.** In the 1990s Motorola created a global telephone system called Iridium. This system was a

combination of 66 low Earth-orbiting satellites able to cover the entire landmass of the world with telephone service. Iridium obtained its own "country code" from the global telephone system, becoming, from a communications perspective, a country. Wal-Mart is another example: its sheer size and geographic distribution dwarf most nations of the world. These companies are well positioned to take greatest advantage of tax-friendly nations in locating their operations, in the same way that many domestic public companies have incorporated in business-friendly Delaware in years past.

- **Virtual services.** Business process outsourcing (BPO) is a growing practice of outsourcing those information functions that can be performed anywhere. Virtualization of information service functions is accelerating, with service providers dispersed globally. Two powerful trends are driving the global information processing market: (1) inexpensive and reliable global digital communications and (2) the increasing number of international workers trained in expanding international universities. An expanding list of job functions can be virtualized globally: software development, electronic design, computer-aided design and drafting, accounting, billings and payables, financial analysis, tax preparation, telemarketing, telephone service support desk, medical record transcription, etc. Any job function is a candidate for virtualization if it can be accomplished with a personal computer, an Internet connection, a software application, and a set of learned rules or procedures (tax code, building code, medical terminology, tech support, accounting, drafting, etc.). With communication costs minimized, work can be transported to the workers; workers need not travel to the work. The long-term economics of competition demand that work be performed in the most cost-effective way. All outputs being essentially equal (the drawing, the tax return, the completed loan application, etc.), work will ultimately find the least expensive labor pool. Since the international labor pool has developed its skills, and the Internet has provided cheap access to those skills, information service functions are being virtualized rapidly.

Technology has crushed economic barriers to international communications among free nations. But ultimately, this system will not support total labor rate differentials of more than a few percent. Currently, international burdened labor rates vary by hundreds of percent. An experienced software developer can be hired in India for less than $12 per hour compared to rates 2 to 4 times that in the U.S.[8] Forrester Research estimates that over 15 million white-collar jobs will depart from the U.S. to the global community in the next 5 to 10 years.

What are the alternatives to these trends, particularly domestically? The most direct, obvious, and painful is an adjustment in labor rates and benefits for competitive endeavors. The less direct, but more likely, scenario is a shift in service job functions to those endeavors that require a "face-to-face" component. As manufacturing employment is giving way to restaurant/retail employment, information processing employment may rebound in the form of personal service endeavors like data entry, training, regulation, inspection, and on-site system support. As in the past, unforeseen innovations will always play an impact role in the demand and type of employment opportunities. The rise of the large middle class in the U.S. in the twentieth century can be largely attributed to Henry Ford's invention of mass production methods, creating a large number of skilled industrial positions. The growing number of professional occupations that served the robust economy has sustained that middle class through the last 50 years.

With the global shift in manufacturing jobs and trends in exporting support functions for business processes, the commercial sector may not be sufficiently robust to support the continued levels of deficit spending for both trade and government budgets. The government has two major economic weapons in its arsenal to combat these techonomic challenges without focusing on the fundamental issue of competitiveness. These two related temporary fixes are currency devaluation and printing of currency. Both ultimately lead to inflation. The tougher, but longer-term answer, is living within national means — balancing budget and trade.

FROM ADAM SMITH TO TECHONOMICS

He (the businessman) *generally, indeed, neither intends to promote the public interest, nor knows how much he is promoting it. By preferring the support of domestic to that of foreign industry, he intends only his own security; and by directing that industry in such a manner as its produce may be of the greatest value, he intends only his own gain, and he is in this, as in many other cases, led by an Invisible Hand to promote an end which was no part of his intention.*

— **Adam Smith,** ***Wealth of Nations***

Adam Smith published his *Inquiry into the Nature and Causes of the Wealth of Nations* in England in 1776, at just about the time when the U.S. was declaring its national independence.[9] It seems far from coincidental that his economic concepts of supply and demand, the invisible hand of the free market, and the benefits of industrious self-interest have been cornerstones of the U.S. capitalistic economic system for the past 200 years.

The genesis of the global economy in the virtual world impacts the fundamentals of Adam Smith's theories. Supplies of "virtual" goods — information — have no production limitations, no distribution limitations, and no demand time lag. The same is true with the "virtual" global labor pool: responsive, distributed, and scalable. The self-interest of the business remains, seeking the lowest-cost provider *de jour* to maximize financial results, no longer favoring domestic vs. foreign industry. A nearly infinite elastic "virtual" supply of information-based products and information workers shifts the rules of marketplace engagement from regional supply and demand to global survival of the most efficient.

Bolstered by technology's glove, the invisible hand guiding organizational evolution now spans the entire globe. Perfect knowledge in the marketplace allows the buyer to optimize the cost–quality techonomic metric for every significant

The Techonomic Future

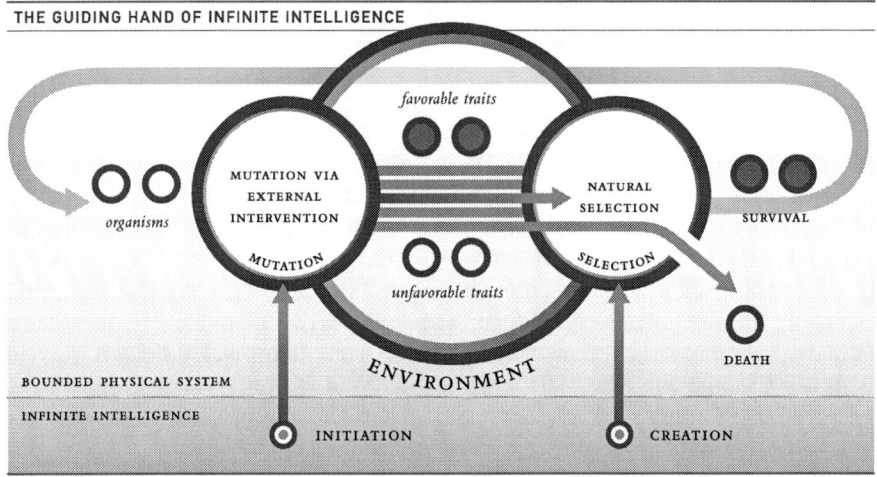

FIGURE 9.1 Concept of system boundaries for Darwin's Theory of Evoluion — the Guiding Hand of Infinite intelligence.

transaction. Winners and losers are not only determined by their cost of production, but by their effective use of technology to manage cash flow, supply chain, promotion, and distribution.

The "self-interest" of the organization is to perpetuate its existence. Wise choices regarding innovation, technology deployment, and marketplace strategy yield results that strengthen the organization. Techonomics postulates that the "Invisible Hand" guiding organizational evolution is none other than the intellect, discernment, and collective experience of the human contributors to, and leaders of, organizations over time. The "Hand" may be diffuse, but it is not entirely invisible, and the cumulative techonomic results are most tangible. ***Figure 9.1 shows the interaction of collective human intellect, the "Guiding Hand," on the Theory of Organizational Evolution.***

The invisible hand, the force of human reason over the ages, can enter the techonomic model in a combination of two ways: (1) through technological advancement or (2) through changing the rules of engagement for the marketplace. The importance of rational and justifiable choices by human intellect to determine the direction of organizations and cultures is more critical than ever before. This is exactly why we need analytical tools like techonomics to hone the decision-making process for effective resource deployment.

THE TECHONOMIC WORLDVIEW

I want to know the thoughts of God ... all the rest are details.

— **Albert Einstein**

Worldview: *noun.* (1) The overall perspective from which one sees and interprets the world. (2) A collection of beliefs about life and the universe held by an individual or a group.[10]

Worldviews are fundamental to the way individuals and cultures analyze and address the events that shape their times. Among other influencers, worldviews vary by nation, culture, religion, ethnicity, generation, educational attainment, and political affiliation. Family, school, friends, media, church, community, public figures, and significant life events shape worldview.

Our worldview filters our observation and understanding of the events around us. If my filter is blue, everything I observe will have a blue hue. If your filter is yellow, yours will be a yellow world. Sounds trivial, but it is not. Your worldview colors everything you see and hence affects your actions in response to those observations. You must actively consider and develop your personal worldview, lest you be tossed by the media-driven winds of "public opinion."

There are political worldviews (Democratic, Republican, Libertarian), social worldviews (conservative, moderate, liberal), religious worldviews (Islam, Christianity, Judaism, atheism, humanism, secularism), governing structural worldviews (democracy, dictatorship, aristocracy), organizational worldviews (bureaucracy, meritocracy), economic worldviews (capitalism, socialism, feudalism), and economic class worldviews (elitism, entitlement mentality, etc.). Knowingly or not, each of us creates an evolving, personal worldview from segments of the foundational principles gleaned for our interactions.

According to Peter Schwartz in *The Art of the Longview,* "Mind-sets [i.e., worldviews] tend to keep us from seeing the appropriate questions to ask about a decision."[11] Conscious and intentional worldviews can be valuable tools for filtering the abundant information experienced in life. Fallacious worldviews created by a façade promoted by powerful ulterior motives can hinder debate and rational decision making in a family, business, or nation. The value of the worldview proposed here — the Techonomic Worldview — is that it provides a process for anticipating future trends and making rational judgments based on factual, statistical, historical information. This view will lead you to solid conclusions about how technology is driving the economy, and thus, how to effectively deploy organizational resources. A purposeful and rational worldview should guide not only our business decisions, but also influence our civic awareness and action.

CAPITALISM VS. SOCIALISM

The twentieth century witnessed an ideological battle between capitalism and socialism. Socialism is "a system of society or group living in which there is no private property; a system or condition of society in which the means of production are owned and controlled by the state."[12] Capitalism is "an economic system characterized by private or corporate ownership of capital goods, by investments that are determined by private decision, and by prices, production, and the distribution of goods that are determined mainly by competition in a free market."[13] The conflict is fundamental and clear. One ideology, socialism, eliminates private property, deferring to the state for deployment of productive resources. The other, capitalism, supports private ownership and expects private entities to determine the production and pricing of commerce. Techonomic theory postulates that organizations evolve within the competitive marketplace, just as biological organisms evolve and survive via the process of natural selection.

Governance in this world ranges from capitalism to socialism — which, in its extreme form, is complete control by the state. Techonomics favors capitalism. Even the economic failure of the USSR, made evident by the collapse of the Berlin Wall in 1989, was not sufficient warning to some nations. Techonomics predicts that such a culture or nation based on economic equality would, at best, assure equal *poverty* for all. Socialist economies cannot successfully compete in the global marketplace with those cultures embracing competition and the best practices that result. It is instructive to observe the economic rise of China as it embraces many tenets of capitalism. Note too, the competitive decline in manufacturing in the U.S. and European Union as they embrace tenets of socialism in key foundational institutions such as healthcare and education.

Free Market vs. Protectionism

The conflict between a free and a protected market arises as large organizations that can no longer compete on a level playing field seek to maintain their standing by regulatory protection. Protectionism typically exists within national boundaries where government market restrictions and tariffs limit the economic competitiveness of imported goods.

Free markets drive production to the lowest-cost provider. As electronic data interchange, GPS package tracking, and automated loading/unloading facilities have reduced shipping and handling costs, it has become more cost effective to send manufactured goods around the globe from the least expensive manufacturer. The dramatic reduction in global communication cost in the last decade is having the same effect on service transactions. The hallmark of the free market is competition, and the salubrious side effect of competition is the adoption of best practices. The hallmark of protectionism, on the other hand, is special interest, and the poisonous manna of special interest is political graft.

Techonomics favors the free market, even when it causes near-term pain while the organization or nation is seeking to redefine its competitive foundations. At best, protectionism only prolongs the steady decline of the organizations it seeks to protect.

Globalism vs. Nationalism

If any transaction in the world is an e-mail away, what is the value and meaning of national sovereignty? Is the European Union an example of future regional socio-economic cooperation, or is it a union in name only, with each nation continuing to work the system to its own benefit? For that matter, is the North American Free Trade Agreement (NAFTA) a treaty for economic benefit of the parties, or does it foreshadow a troubled merging of the nations, the cultures, and the people, as the chaos of the U.S. southern boarder immigration situation suggests? These are tough questions without clear answers.

Techonomic theory returns to the importance of competition as a determinant of continuous improvement. People and organizations, by their fundamental nature, are competitive. Nations, like companies, learn from competition. Best practices in government often result from implementing a successful approach observed in another place. A one-world-order approach will tend toward concentration of tremendous power in the hands of a few — and a level of equalized mediocrity for the masses.

By eliminating international economic barriers to trade (primarily through minimizing communications costs), the advance of technology is hastening the effect known as the Law of the Mean. Standards of living for the worldwide labor force will trend toward a global standard *in countries where cooperative people work in a productive framework.* The local integrity of sovereign governments will be fundamental to the well-being of citizens.

Personal Responsibility vs. Entitlement

Cumulative advances in quality of life due to technology gain have provided the vast majority of U.S. citizens with access to food, entertainment, information, education, transportation, clothing, and shelter surpassing the wealthiest inhabitants on Earth 200 years ago. Unfortunately, this observation does not equate to universal happiness, because significant disparity in living standards remains. The significant differences in the distribution of wealth between groups and between nations that coexist in this materialistic age will be an eternal source of struggle. Any organization or nation that does not perpetuate an expectation of *personal responsibility* for "life, liberty and the pursuit of happiness" will ultimately collapse under the weight of its own national "prosperity," resulting in mediocrity. We must use our educational system and social policies to reinforce the values of personal responsibility and self-reliance. The land of plenty must not become the home of the dissatisfied.

There is a danger inherent in improvement based on technological advance. As our world becomes increasingly dependent on technology, we will become increasingly specialized in our professional endeavors and increasingly dependent upon others for life's necessities and comforts. Active, cooperative *interdependence* is the road to health; entitlement-sustained *dependence* leads to personal inactivity and organizational devolution.

Merit vs. Diversity

From the 1970s to the bursting of the Internet Bubble, the spirit of Silicon Valley exemplified and accelerated techonomics. The rise of the meritocracy (an organization where promotion was based on merit) brought with it the creation of new products, services, and wealth at rates humanity has never before seen. People from all parts of the globe cooperated, pushing the boundaries of science to advance economic prosperity for the region — and in many ways for the nation and the world.

A meritocracy characterized the corporate structure that emerged within many of the hundreds of startups. Talented and highly valued contributors were retained by equity and options to keep them from seeking elsewhere for better economic returns on their services. These corporate melting pots were an amalgam of diverse intellect drawn from the four corners of the world to seek the greatest challenge and opportunity of the time. The key lesson to be learned from the technologic and economic miracle of Silicon Valley was the productivity that resulted from merit-based organizational structures within a highly competitive, free market economy. Diversity in the absence of merit paves a pathway to decline. Merit without diversity will be blindsided by myopia.

The Invisible Hand Leading Organizational Evolution

Where do the organizational trends anticipated by techonomic theory lead? Techonomics predicts survival of the fittest organizations in the form of continuously greater efficiency and productivity. Human ingenuity will continue to extend the boundaries of knowledge and harness its secrets for productive purposes. The creative human imagination will dream and produce technologies and marketplace methods that will improve the collaboration and endeavors of organizations. New organizational structures will evolve under the invisible hand of cooperative people and succeed in the marketplace. As shown in Figure 9.1, the external invisible hand moving organizations and productivity forward is nothing less than the collect output of the human mind. Innovations and improved structures will proliferate through organizations at an increasing rate as information on best practices becomes rapidly diffused due to improving and growing communications networks. On the crest of techonomic trends, the organization of the future — the virtual corporation — will do more and more with less and less until it is able to do very much with very little.

We are headed rapidly into the Virtual Age, where the capability and ubiquity of our technological creations will enable us to accomplish and experience most endeavors in life without need for physical presence. Convergence of the "-tions" — information, communication, digitization, integration, computation, immersion, simulation, automation — is imminent. Convergence in the Virtual Age encompasses the physical and technical infrastructure to support a virtual existence.

The science fiction movie *The Matrix* takes this concept to the big screen, where the virtual and the real are indistinguishable and the body becomes simply a container for the mind. While the Matrix remains science fiction, the advances of productivity and virtual extension predicted by the twenty-first-century Techonomic Laws pose a very individualistic, independent future. A conflict is growing between the Law of Increasing Productivity, requiring a smaller workforce to accomplish more output, and the fundamental techonomic metric for community: growth. This conflict will mark the decades ahead as the U.S. strives for greater productivity while maintaining living standards for the mass populace. The trend toward specialization will lead to increasing interdependence, and basic life skills of the greater community will diminish unless these skills are carefully preserved. We are increasingly dependent on technology to sustain life (electricity, logistics support, communications, etc.).

The technology-enhanced quality of life will be limited only by the imagination, as virtual experiences allow us all to realize the imaginable.

In order to grasp the full possibilities of this future, we must exercise the collective will to preserve the foundations of competition in all markets, most particularly the three industries in crisis: energy, education, and healthcare. The wise stewardship of our natural and economic resources requires a significant course correction in these industries if the next generation is to have the resources necessary to harness the innovative minds and entrepreneurial spirit that built this nation's prosperity.

May the *Theory of Techonomics* provide you with clear, actionable insights into the workings of twenty-first-century organizations. Use your resources well, like the good stewards who multiplied the wealth of their master and were then put in charge of managing more wealth and greater organizations.[14]

SUMMARY

- Technology mutates how organizations operate.
- The economy is the environment in which organizations must exist.
- Free market competition causes organizational natural selection.
- Techonomic progress mandates survival of the most efficient.
- Human intellect is the invisible hand stimulating techonomic evolution.

TECHONOMICS' INVISIBLE HAND

The "Invisible Hand" spoken of by Adam Smith was the cumulative effort of a myriad of individuals, all acting in their own best interest, resulting in economic progress. The progress of techonomics toward increasing productive efficiency is motivated by the intellect of people leading organizations within a competitive economy. Good technology resource deployment leads to favorable economic results. Enough poor technology resource deployment leads to bankruptcy. Human intellect serves as the external impetus for productive organizational progress monitored and anticipated by techonomics.

TECHONOMIC EVOLUTION

Techonomic evolution has an obvious external stimulus (human intellect). Leaders have driven progress through innovative technologies and economic models, while followers have embraced leaders by adopting their best practices. The organizations that have done this most efficiently have grown and prospered. Through the years, the result has been a march toward greater productive use of all resources.

THE VIRTUAL AGE

As humankind accomplishes increasingly more tasks remotely, we have moved beyond the agricultural age, the industrial age, the space age, and the information age into the *Virtual Age*. This age will be marked by greater specialization, greater interdependence, and greater life-experiences without physical presence.

Key Terms

> virtualization
> business process outsourcing
> macro-multinational
> worldview
> Law of the Mean

QUESTIONS

1. Adam Smith states that *"he* [i.e., the worker] *intends only his own security."* Do you agree with Smith's assertion that economics works because individuals act in their own self-interest? Why or why not? If the rewards of self-interest were removed from the economy, or from a major segment of it, what would Adam Smith's view predict?
2. Techonomic theory proposes that an organization acts in its own interest to perpetuate its existence to the benefit of its constituents (workforce, shareholders, management, customers). If an economic or regulatory structure is established that removes or reduces the organization's performance incentive, what techonomic predictions would you make concerning the long-term productivity of the organization?
3. Some people fear that an increasing virtual dimension of life will diminish rather than enrich the quality of life. Do you think this is true? How so? Try making a list of both the benefits and the dangers of the Virtual Age as you envision it. Which list is longer? Are the benefits worth the risk of facing the dangers? Are the dangers avoidable? How?
4. Other than human intellect guiding organizations, what other forces would you suggest act as "Invisible Hands" of organizational evolution? (Hints: capital markets, economic structure, cultural work ethic.) Are these significant stimuli relative to human intellect?
5. The Virtual Age predicts that major opportunities are to be made by "virtualizing" activities that are currently done with direct physical interaction. Think of an activity or transaction that you do currently that could be accomplished remotely. List the activity and describe methods that could be used to perform the activity virtually.
6. The national debt increases when the Federal budget is not balanced (expenditures exceeding revenues). To finance the debt, the government typically sells 30-year Treasury notes on a weekly basis. The present value of past debt decreases each year at the rate of inflation ($20,000 debt in 1960 would purchase a nice house; $20,000 debt in 2000 would purchase a nice automobile!) Can you conceive of the conditions necessary to remain spending in deficit perpetually? What role does inflation play in your scenario? What role does the ability to print money play in your scenario?

REFERENCES

1. Zebra Imaging, Holographic images, Zebra Imaging, 2005, http://www.zebraimaging.com/html.
2. Merkle, R.C., Nanotechnology, http://www.zyvex.com/nano/.
3. Sondergard, M., A doctor's prescription: swallow one camera-in-a-pill and call me in the morning, University of Iowa Health Care, University of Iowa, 2004, http://www.uihealthcare.com/news/pacemaker/2004/summer/camerainapill.html.
4. Walter, K., A new precision cutting tool: the femtosecond laser, Lawrence Livermore National Laboratory, University of California, 2006, http://www.llnl.gov/str/Stuart.html.
5. Plasmas International, Perspectives on plasmas, basics, 2004, http://www.plasmas.org/basics.htm.
6. Eck, J., Type 2 Superconductors, Superconductors.org, 2006, http://www.superconductors.org/type2.htm.
7. Eck, J., Superconductors, Superconductors.org, http://www.superconductors.org.
8. Mapsofindia.com, Job outsourcing to India, Maps of India, Mapsofindia.com, http://www.mapsofindia.com/outsourcing-to-india/index.html.
9. Smith, A., *Wealth of Nations,* Great Mind Series, Prometheus Books, New York, 1991.
10. *The American Heritage Dictionary of the English Language,* 4th ed., Houghton Mifflin Company, 2000, The Free Dictionary, http://www.thefreedictionary.com/worldview.
11. Schwartz, P., *The Art of the Long View: Planning for the Future in an Uncertain World,* Doubleday Books, New York, 1996.
12. *Merriam-Webster Online Dictionary,* Merriam-Webster Inc., 2005, Merriam-Webster Online, http://www.m-w.com/dictionary/socialism.
13. *Merriam-Webster Online Dictionary,* Merriam-Webster Inc., 2005, Merriam-Webster Online, http://www.m-w.com/dictionary/capitalism.
14. Matthew25: 14–23; Luke 19:17.

Afterword

> But by acting according to the dictates of our moral faculties, we necessarily pursue the most effectual means for promoting the happiness of mankind, and may therefore be said, in some sense, to co-operate with the Deity, and to advance as far as in our power the plan of Providence. By acting other ways, on the contrary, we seem to obstruct, in some measure, the scheme which the Author of nature has established for the happiness and perfection of the world, and to declare ourselves, if I may say so, in some measure the enemies of God.
>
> — **Adam Smith,** *The Theory of Moral Sentiments,* **1759**

Throughout *Techonomics,* I have made reference to the father of modern economic theory, Adam Smith — especially to his "Invisible Hand" analogy. Smith's ideas are the fountainhead of free market economic thought. But it is important to remember that Smith wrote another seminal work entitled *The Theory of Moral Sentiments* (1759).[1] In this book, Smith says the root of healthy human behavior is a God-implanted sympathy for other human beings. It is difficult to write about economics and human development without engaging the deep questions of human nature, human responsibility, and even human origin.

As I wrote *Techonomics,* the numerous parallels between organic and organizational evolution unfolded in my mind. I am aware that I am not the first to use a structure reflecting the theory of organic evolution as a basis to understand other phenomena. Still, it came to me as something of a revelation when I perceived the close fit between Darwin's central idea and the development of human organizations through the ages. Intelligent application of technology, operating in free marketplaces (the economic environment) had become the primary "favorable mutation" through competitive "natural selection" that allowed one organization to prosper while another faded away. It is something I could track historically, see currently, and predict its ramifications confidently.

Techonomics provides an observable structure for organizational evolution in past and current organizations. We can observe techonomic trends, their implications, and their adaptation by intelligent human beings. But as for the ultimate questions — what Invisible Hand may be supporting human development overall, and what Hand started the process of organic evolution — these cannot be answered by direct observation.

The debate rages about Intelligent Design as an alternative to Darwin's theories explaining the origin and diversity of life. Not one scientist or theologian was present at the Big Bang. No one had the privilege of observing a repeatable event, and it cannot be recreated mathematically without a myriad of truly unobservable assumptions. Science requires observation. What cannot be observed can be accepted or

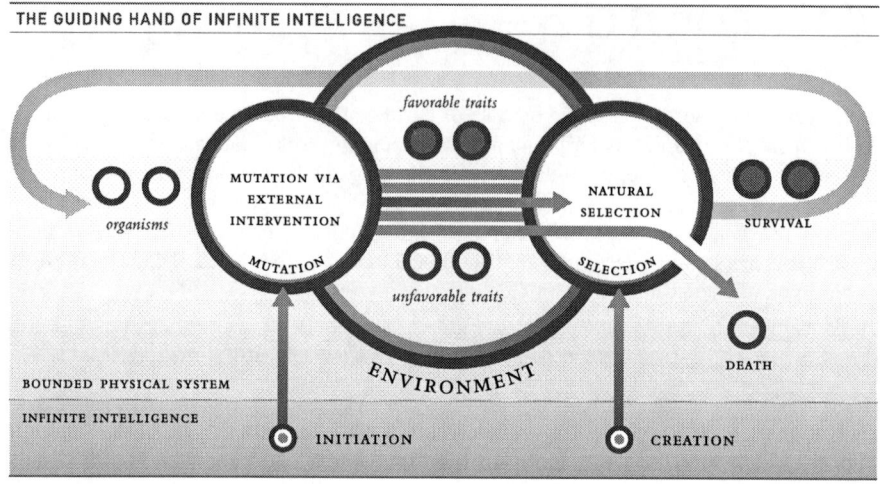

FIGURE 9.2 Concept of system boundaries for Darwin's Theory of Evoluion — the Guiding Hand of Infinite intelligence.

rejected only by faith — faith in either Science as Measure to All Truth, or in the Creator-God as the Intelligent Designer.

Each of us has personally and privately pondered the source of life's creation and development. For me, having observed the role of human intelligence in guiding the process of organizational evolution, it is a logical step to conclude that an active Creator initiated and guides the process of organic evolution. A Creator initiating, guiding, and shaping the organic world seems in some way to be reflected in the intelligent beings He made in His image, and who have innovated and shaped the organizational world. *Figure 9.2 shows the system boundaries for Darwin's Theory of Evolution with consideration of the Guiding Hand of an Intelligent Designer.* Human intellect has been the constant influence guiding organizational evolution throughout recorded time. With so many other parallels between the two theories described herein, I would infer that the Creator has been the external influence guiding the development of organic evolution throughout eternity.

Faith, whether placed in Darwin or in the Creator, should not be a subject for distain, educational exclusion, or condescension on either side. Faith, along with moral convictions and intellect, is fundamental to binding individuals into the greater, cooperative community. It is faith that forms our goodness. In the profound words of Alexis de Tocqueville, "America is great because America is good; and if America ever ceases to be good, America will cease to be great."[2]

— **Lee Martin**
February 2006

REFERENCES

1. Smith, A., *The Theory of the Moral Sentiments (1759),* Adam Smith Institute, 2001, http://www.adamsmith.org/smith/tms-intro.htm.
2. Federer, W., *America's God and Country: Encyclopedia of Quotations,* Fame Publishing, Inc., Coppell, 1994, p. 205.

Appendix 1
Terminology Related to Techonomics

Acceptable Risk. Every activity carries some degree of risk for economic or physical injury. Acceptable risk defines the severity and probability of injury an organization is willing to tolerate for a given activity in order to receive the benefits offered by that activity. No risk, no reward. Too much risk, too much liability.

Automated Bureaucracy. This is a system in which policies and procedures have been implemented with intelligent machines rather than human labor.

Best Practices. The most cost- or labor-effective approaches to endeavors developed by an industry. Smart competitors learn from others and often embrace best practices to improve their own operations.

Burdened Labor Rate. The total cost of human labor including wage rate, all benefits costs (health insurance, retirement plan, etc.), and employment taxes (Social Security, unemployment tax, workman's compensation, etc.). Burdened labor rates can exceed wages by 50 to 100% in the hidden economy of mandated benefits.

Business Process Outsourcing (BPO). The systematic development of a provider network to perform the basic activities associated with fundamental business operations: accounting, distribution, record keeping, etc. BPO is a subset of outsourcing, referring to the information-based activities fundamental to business operations.

Capital. Money, particularly as used to fund the productive capacity of organizations.

Capital Formation. The process of acquiring capital for an organization. Capital formation is easier in markets with a track record of return on investment and consistent legal policies.

Chained Dollar. A measure of inflation based on annual purchase of an equivalent "basket" of household goods. Instituted in 1996 to replace the inflation-adjusted dollar, the chained dollar benefits from a built-in electronics cost reduction resulting from the effects of Moore's Law.

Coase-Downes-Mui Law. The Law of Increasing Productivity results from considering Coase's transaction analysis along with the effect of "perfect information" noted by Downes and Mui.

Comparative Techonomic Metric. A measurement combining technical performance and cost to compare alternative means of performing the same endeavor. Comparison might be between different manufacturing plants, different materials, or different processes all producing the same product. The goal is to determine the best approach (find the "best practice").

Competition. A force resulting from the free market in which multiple organizations perform the same, or very similar, endeavors for commerce. Competition is the impetus behind techonomic evolution: it causes best practices to be developed and embraced, leading to greater productivity.

Consumer Price Index. A measurement of inflation, annually adjusted, used as a benchmark for the condition of the economic system.

Continuous Improvement. An operational philosophy seeking to promote progress and productivity by constantly advancing operating techniques based on innovation and incorporation of marketplace best practices.

Cost-of-Living Adjustment. A financial adjustment in entitlement and employment agreements that increases these payments every year based upon changes in the Consumer Price Index or other accepted measures of inflation.

Customer Labor Component. In an effort to minimize labor costs, producers have systematically automated manufacture and shifted some labor components to the customer ("some assembly required"). Franchises are exceptional at this, with customers responsible for checking out, serving their own food, and bussing their own tables.

Digital Omniscience. A hypothetical condition (but one drawing nearer to reality all the time) in which communications networks become so pervasive, and computationally capable, that networked computers know everything about everyone.

Earnings. Profits an organization receives for performing an endeavor (basically, revenues − expenses).

Endeavor. The output of the organization: products, services, or other activities that create exchangeable value. Transactions spring from the exchange of endeavors between organizations.

Endeavor Cash Flow. The measurement of capital and its time value use in the process of performing an endeavor.

Franchise Effect. Franchising is a method of organizational growth that consists of replication of successful "units." Local operational models are perfected and multiplied regionally, nationally, and internationally to grow an organization, while operations in each new location are kept small and efficient.

Free Cash Flow. At the completion of an endeavor, this value is the revenue minus the cost of production including any capital expenses (facilities and equipment) necessary for production. Free cash flow does not include consideration for the time value of money as endeavor cash flow does.

Free Market. Free market commerce is characterized by multiple suppliers, domestic or international, with equal opportunity to transact business.

Terminology Related to Techonomics 211

Frictionless Economy. "Frictionless" economy refers to an ideal condition where perfect information about all supply source possibilities is readily available. A frictionless economy enjoys abundant information and no procurement boundaries (tariffs, prohibitions, etc.).

Globalization. Globalization refers to the increasing connectedness of the world's markets, businesses, and production networks. This increase in the distributed nature of production results directly from the three primary laws of techonomics: increasing computational capability, increasing communication connectivity, and increasing productivity of organizations.

Gross Domestic Product (GDP). This value is the sum of the economic value of all goods and services produced annually by a nation.

Hand-Held Convergence. The merging of all communications and media devices into a single, portable, wireless unit including cell phone, music, radio, television, Web access, photography, videography, and e-mail.

Hedonic Correction. An adjustment to the GDP based on the utility of goods produced rather than the actual money generated from the exchange of those goods. Due to Moore's Law, this correction accounts for an ever-increasing contribution to the GNP by technology advance that is not substantiated by cash flow from purchases.

Hidden Costs. Hidden costs are all the expenditures for a transaction in addition to the actual price. These include costs of switching, finding, rejection, rework, disruption, etc.

Historic Techonomic Metric. A measurement comparing the current technology output per unit measure relative to the same production at different points in time. Because of the limited timeframe for currency relevance, these measurements typically must use an economic component other than dollars. Production measurements relative to labor hours, land use, material consumption, etc. form the basis for historical comparison of techonomic progress.

Inflation-Adjusted Dollar. A traditional measure of inflation based on yearly purchase of basic commodities without correction for technology advance.

Interdependency. This term is the opposite of self-sufficiency. This is the concept of community wherein all members produce and exchange endeavors with others in order to survive and prosper.

Just-in-Time Delivery. JIT is a manufacturing process where supplies are provided only when needed, eliminating large inventories, storage facilities, or extensive amounts of work in progress.

Law of Increasing Productivity. This term is synonymous with the Coase-Downes-Mui Law and the Third Law of Techonomics. This Law predicts continuing improvement in productivity due to transaction analysis powered by perfect information.

Law of the Mean. All things tend toward an average unless external forces maintain a difference.

Law of the Ubiquitous Global Network. Synonymous with Metcalf's Law and the Second Law of Techonomics, this Law predicts the World Wide Web will continue to grow until every person, and most objects, become addressable on a wireless, ever-present network.

Law of Ubiquitous Computing. Synonymous with Moore's Law and the First Law of Techonomics, this law predicts incorporation of intelligent microcomputers into an increasing array of products as microprocessor cost diminishes exponentially. Virtually every product will be "smart" and getting smarter all the time. Human intellect will add less value to the bureaucratic or "standardized" workplace, and more to pure research, technological innovation, and creativity of all sorts, as machine intelligence automates more labor.

Lean Organization. A business that has trimmed most of its organizational overhead by outsourcing support functions to the most effective supplier.

Macro-Multinational. This term refers to a very large multinational firm that exhibits the characteristics of a nation because of the extent of its operation, revenues, and global influence.

Make-or-Buy Decision. In transaction analysis, a layman's term referring to the decision about the most effective way to carry out an endeavor. Basically: Shall I do task X myself or hire someone to do it for me, so I can concentrate my efforts and resources elsewhere?

Mass Customization. A manufacturing process wherein the customer chooses the configuration of a sophisticated final product, and the manufacturer makes it to order. Twenty-first-century business networks and electronic data interchange make this mode of operation possible on a scale never before conceivable.

Media Convergence. The merging of all entertainment information into a single delivery platform: television, music, video games, Internet, etc.

Metcalf's Law. Named for the observation of Bob Metcalf, founder of 3Com, this law is the mathematical observation that the connections on a network grow exponentially relative to the number of users. (Connections = $0.5 \times (X^2 - X)$, where X is the number of nodes.)

Metric. A measurement developed from available data to monitor a key performance trend.

Micro-Multinational. This term refers to a start-up company with a core leadership team, typically in the U.S., leveraging international labor for development and manufacture in order to minimize the capital needed for starting and expanding operations.

Moore's Law. Named for Gordon Moore after an observation he made in a 1965 technical paper. He stated that electronic transistors would be packaged twice as densely every 18 months, hence reducing their costs or increasing their performance proportionately.

Organic Evolution. A theory stating that living organisms sprang from a common ancestor or a few ancestors and, over the eons, morphed into the diverse life forms observable today. This occurred through a continuing process of mutation and natural selection (the environment favoring survival and reproduction in organisms possessing even slight mutations/differences that helped them in some way: slightly better eyesight, speed, coloration, etc.). As used here, the term is synonymous with Darwin's Theory of Organic Evolution.

Organizational Evolution. A theory about the progress of organizations that states they advance in a process similar to the one proposed by Darwin for organisms. Technology causes mutation; the economy serves as the environment; competition provides the filter for natural selection; adaptive organizations grow and prosper; dysfunctional organizations die. Over time, organizations become more efficient and productive as they evolve to improved operational structures. In this book, the term is used synonymously with the Techonomic Theory of Organizational Evolution.

Organizational Span of Control. This is based on the number of individuals reporting directly to the next level of organizational management. The span of control is low (< 7) in a vertically integrated and traditionally managed organization. The span of control is large in a "flat" organizational structure where personal control is augmented by electronic communication and performance management.

Outsourcing. Transferring endeavors from internal performance to an external supplier. With "perfect" information, the impediments to procuring goods from external sources are minimized or eliminated, leading organizations to more purchasing (external) rather than internal production.

Natural Selection. The process wherein organisms compete for life resources in order to survive and are favored in their environments, or not, due to their characteristics (see Theory of Organic Evolution). Natural selection, also known as the survival of the fittest, plays a pivotal role by winnowing out organisms that are unable to successfully compete, leaving the stronger organisms to procreate and produce the next generation.

Planned Obsolescence. A design technique used to estimate the useful life of a product and create the design specifications, and hence the end product, around the desired useful life of the product.

Perfect Information. Absolutely perfect information for making decisions would be free, complete, accurate, timely, and actionable data. This is an ideal, but the closer the approach to this information, the higher the confidence in decisions. Perfect information favors outsourcing in the make-or-buy decision.

Performance/Cost Ratio. Techonomic metrics combine elements of technology measure with economic considerations. Performance/cost ratios can be used for comparing stock values, manufacturing processes, resource use, etc.

Positive Cash Flow. The time difference between receipt of customer's money and requirement to pay suppliers can provide money to an organization. If an organization's endeavor is sale of computers, for example, that organization may receive a customer's money months before payment is due to vendors who supplied the computer's components. Online ordering, mass customization, and favorable supplier terms lead to enhancing positive cash flow.

Predictable Antiquation. A shortened useful life of many electronics products is the result of continuous improvements in performance enhancement resulting from Moore's Law. This advance causes a computer bought today to be obsolete in 2 to 5 years, even though it remains fully functional. It is no longer compatible with the software and peripherals emerging in the industry and the user may demand faster system performance.

Product Life Cycle. The life cycle of a product refers to the period of time in the marketplace that a particular design will remain a competitive offering. Due to the impact of Moore's Law, the product life cycle for consumer electronics is typically less than 2 years.

Renewable Energy. An energy source that can be replenished regularly is referred to as renewable. This term is used to refer to solar, wind, hydro, wave, bio-fuels, and others that regenerate cyclically through the day or season. Nuclear breeder reactions also create more fuel than they consume, but are seldom considered in the definition of renewable energy.

Rolling Warehouse. Just-in-time delivery often uses trucks for delivering supplies to product integrators. The work in progress is the work inside the truck, hence the name rolling warehouse.

Self-Interest. It is the nature of all organisms and organizations to preserve their own lives. Self-interest refers to the motivation to make choices conducive to survival.

Switching Costs. In transaction analysis, the switching cost is the cost or risk associated with replacing a supplier should a problem arise. If alternative suppliers are plentiful and outages are not critical, switching costs are low. When there are no known alternatives, or the cost of downtime is great, switching costs are very high.

Techonomics. Techonomics is a theory of organizational evolution resulting from the study of technology trends and their economic effects on organizations.

Techonomic Metric. A measurement based on technology performance and economic cost. It is used to anticipate trends in organizations.

Techonomic Sweet Spot. The combination of performance advantages and cost savings yields a target where new technology offers significant market positioning opportunities.

Transaction. An exchange between parties of endeavor for monetary compensation is the essence of transactions.

Transaction Costs. The expenses associated with transactions that determine the total economic consideration for an exchange. In addition to the price itself, transaction costs include the cost of transportation, maintenance, replacement, quality assurance, vendor reliability, etc.

Virtual Age. The Virtual Age is marked by an ever-expanding digital infrastructure capable of fulfilling an increasing array of life needs. We have passed through an Agricultural Age and an Industrial Age, and more recently through the Atomic Age, Space Age, Information Age, and Biotechnology Age. The Virtual Age is based upon the three current trends of techonomics that will propel organizational change throughout the first half of this century.

Virtual Company. A virtual company is structured around outsourcing all possible tasks that contribute to an endeavor. A virtual company may outsource manufacturing, fulfillment, retailing, distribution, support, service, etc. — leaving only a small cadre of leaders to orchestrate the activities of the organization.

Vertical Integration. Vertical integration is a productive system wherein a significant portion of a particular endeavor is produced by one organization.

Virtual Retailing/Storefront. A virtual retailer retains no brick-and-mortar outlets, but chooses to sell products and services virtually via networks like the World Wide Web.

Virtualization. Moving real, tangible activities into the realm of the simulated and remote is the process of virtualization.

Worldview. Worldview is the overall perspective from which one sees and interprets the activities of the world.

Zero Risk Tolerance. This philosophy is guided by the premise that there are no acceptable societal risks, no matter what the societal benefit. Applied to transportation, for example, it would have prevented acceptance of the automobile (or bicycle, for that matter). Applied to healthcare, it would have prevented acceptance of aspirin or any other pharmaceutical that followed. Intelligent, responsible acceptance of some risks is fundamental to technological advance and economic wellbeing. The "zero risk tolerance" attitude is widespread, misguided, and unrealistic. It is decision making based on magnification of fear and will stymie those societies caught in its grasp.

Appendix 2

Example of Process for Developing a Techonomic Metric

1. Select a market and time frame to study.
   - *Digital cameras, 1980 to present.*
2. Select the key technology component to be measured.
   - Pixels in the imaging chip for the camera.
3. Select the key economic component to be measured (typically dollars, but it may be labor hours of content or materials use).
   - Retail cost for complete high-end consumer camera. ($1000 circa 1996).
4. Gather technical and economic data for time frame of interest.
   - Historic configurations and prices, found on Web and in product literature.
   (~Resolution = 1,300,000 pixels circa 1996).
5. Combine the technical and economic components into a single metric.
   - Techonomic metric defined by pixels/$.
   (1,300,000 p/$1000 = 1300 pixels/$ circa 1996).
   - Techonomic metric calculated using data available at various points in the history of the market.
6. Combine data and plot it, looking for trends shown by the slope of the techonomic metric curve.
   - See Figure 4.2.

Index

A

Agriculture, techonomic metrics, 73–74
    productivity gains, 74
Amazon, as business model, 120–123
American Telephone and Telegraph, breakup of, 149
Antiquation, predictable, 116–119
Apple, as business model, 125–127
Assumptions in techonomics theory, 12–13
    competition, 13
    free market, 13
    knowledge expansion, 13
    self-interest, 13
Australia, student performance in, 169
Austria, student performance in, 169
Automated bureaucracy, 147
Availability, hidden transaction cost, 21

B

Belgium, student performance in, 169
Bell System, breakup of, 149
Bhabha, Dr. Homi, 54
Biological energy sources, 134
Biological life, organic evolution theory, 6–10, 16
Biological processes, 191–192
Brodie, Leo, 93
Bureaucracy, automated, 147
Business models, 103–132
    Amazon, 120–123
    antiquation, predictable, 116–119
    Apple, 125–127
    debtless facility expansion, 114–116
    Dell, 104–109
    distribution, positive cash-flow retail, 109–114
    eBay, 123–125
    Intel, 116–119
    manufacturing, positive cash-flow retail, 104–109
    Microsoft, 119
    speed business, 119
    virtual media, 125–127
    virtual reselling, 123–125
    virtual retail, 120–123
    Wal-Mart, 109–114
    Walgreens, 114–116

C

Canada, student performance in, 169
Chemical energy sources, 136–137
China, student performance in, 169
Coase, Ronald, 19–20. *See also* Coase-Downes-Mui law
Coase-Downes-Mui law, 87–88, 93–96, 99, 194
Communications
    emerging trends, 142–147, 152
    evolution in, 39–40
    techonomic metrics, 64–70, 81
Community
    emerging trends, 147–152
    evolution in, 40–41
    techonomic metrics, 70–79, 81
Community health, techonomic metrics, 74–77
Competition, 157–160
    as fundamental assumptions in techonomics theory, 13
Completeness, importance of, 23
Computation
    emerging techonomic trends, 140–142, 152
        digital omniscience, 141–142
        virtual world, 142
    evolution in, 39
    techonomic metrics, 57–63, 81
Computer graphic representation, 186
Computing, ubiquitous, law of. *See* Moore's law
Constant techonomic metric measure, techonomic metrics, cost *vs.* performance, 47
Consumer Price Index, inflation rate, 175
Contemporary energy technologies, 138–140
Cost analysis of transactions, 19–21, 26
    availability, 21
    hidden transaction costs, 21–22
    inventory, 21
    punctuality, 21
    quality, 21
    risk, 22
    switching, 22
    transport, 21
Crises, market, recommendations, 157–182

Cultural approaches to endeavor, techonomic metrics, 45
Customization, mass, 95
Cycle of learning, 4
Cyclical energy sources, 134–136
Czech. Republic, student performance in, 169

D

Darwin's Theory of Organic Evolution, 1–10, 16, 197, 206
de Tocqueville, Alexis, 174
Debtless facility expansion, 114–116
Definition of techonomics, 3, 10–12
Dell, 104–109
Denmark, student performance in, 169
Derivation of term "techonomics," 10–11
Digital photography techonomic metric, 48–51
   camera development, 50
   camera sales, 50
Diminishing organization size, law of. *See* Coase-Downes-Mui law
Distribution, positive cash-flow retail, 109–114
Drug treatments for disease, mass-customization of, 191

E

eBay, 123–125
Economic protectionism, 148–150
Economics, derivation of term, 11
Education, crisis in, recommendations, 158, 167–174, 180
Einstein, Albert, 87, 197
Electrical transmission lines, superconducting, 190
Electronic cost, exponentially diminishing, law of, 194
Embossing, molecular, 188
Emerging trends, 133–156
   American Telephone and Telegraph, breakup of, 149
   communications, 142–147, 152
   community, 147–152
   computation, 140–142, 152
      digital omniscience, 141–142
      virtual world, 142
   entitlement, 150–151
   International Business Machines, paring down, 149
   Microsoft, decoupling of key products, 149
   renewable energy resources, 133–140, 151–152

      biological sources, 134
      chemical sources, 136–137
      contemporary energy technologies, 138–140
      cyclical sources, 134–136
      hybrid vehicles, 139
      nuclear power, 138–139
      nuclear sources, 137–138
      solar-voltaic electricity, 139
      solar voltaic sources, 135–136
      wave sources, 135
      wind sources, 135
   specialization, 147–152
Endeavors, defined, 12
Energy. *See also* Renewable energy resources
   consumption, per capita, 55
   crisis in, recommendations, 158, 160–163, 179
   evolution in, 37–38
   storage devices, superconducting, 191
   techonomic metrics, 54–57, 81
Entertainment, three-dimensional, 185–187
Entitlement, 150–151
Environment, 74–77
Evangelism, virtual, 193
Evolutionary frameworks, parallels, 9
Exponentially diminishing electronic cost, law of, 194

F

Femptoseconds, 189
Finland, student performance in, 169
Four-square principle, 36–41
   for individuals, 41
   for organizations, 42
France, student performance in, 169
Franchise effect, 96–99
Franklin, Benjamin, 174
Free market, as fundamental assumptions in techonomics theory, 13
Fundamental assumptions in techonomics theory, 12–13
   competition, 13
   free market, 13
   knowledge expansion, 13
   self-interest, 13
Future developments, 183–204

G

Generators, superconducting, 190
Genetically engineered plants, 185

Index

Germany, student performance in, 169
Global network, ubiquitous, law of.
    See Metcalfe's law
Globalization, 95
Government, crisis in, recommendations, 174–180
Greece, student performance in, 169

H

Healthcare
    community, techonomic metrics, 74–77
    crisis in, recommendations, 158, 163–167, 179
    expenditures, 76
Hedonic Adjustment, 176
Hidden transaction costs, 21–22
    availability, 21
    inventory, 21
    punctuality, 21
    quality, 21
    risk, 22
    switching, 22
    transport, 21
Historical perspectives, 29–84
    organizational evolution from technology, 31–44
    technological breakthroughs, 33
    techonomic metrics, 45–84
Holographic projections, 185
Hong Kong, student performance in, 169
Human evolution
    four-square principle for, 37
    organizational evolution, technological advancement, 31–44
        communication, evolution in, 39–40
        community, evolution in, 40–41
        computation, evolution in, 39
        energy, evolution in, 37–38
        four-square principle, 36–41
            for individuals, 41
            for organizations, 42
        history of technological breakthroughs, from prehistory to 1500 AD, 33
        human evolution, four-square principle for, 37
        mental evolution, 39
        organizational evolution
            four-square principal for, 38
            timeline, 32–36, 41
        physical evolution, 37–38
        social evolution, 39–40
        spiritual evolution, 40–41
        technology timeline, 41

Human life cycle, organizational life cycle, compared, 7
Hungary, student performance in, 169
Hybrid vehicles, 139

I

Iceland, student performance in, 169
Intel, 116–119
Interdependence, 79–81
International Business Machines, paring down, 149
Inventory, hidden transaction cost, 21
Ireland, student performance in, 169
Italy, student performance in, 169

J

Japan, student performance in, 169
Julian, W.J., 167
Just in time manufacturing/delivery, 95

K

Knowledge expansion, as fundamental assumptions in techonomics theory, 13
Korea, student performance in, 169

L

Latvia, student performance in, 169
Laws of twenty-first century techonomics, 87–102
    Coase-Downes-Mui law, 87–88, 93–96, 99
    franchise effect, 96–99
    Metcalfe's law, 87, 91–93, 99
    Moore's law, 87–91, 98
Lean organization, 95
Learning, cycle of, 4
Liechenstein, student performance in, 169
Luxenburg, student performance in, 169

M

Macro-multinational virtual company, 194–195
Magnets, superconducting, 190
Make-or-buy decision, transaction cost analysis, 20–21
Manufacturing, positive cash-flow retail, 104–109
Market crises, 157–182
    competition, 157–160

Consumer Price Index, inflation rate, 175
  education, 158, 167–174, 180
  energy, 158, 160–163, 179
  government, 174–180
  healthcare, 158, 163–167, 179
  Hedonic Adjustment, 176
  Moore's law, 175
  Organization for Economic Cooperation and Development-Program for International Student-Assessment, 168–171
Martin, Lee, 206
Mass customization, 95
Material treatments, plasma, 190
Media
  influence of, 77–79
  techonomic metrics, 77–79
Media convergence, 145–146
Mental evolution, 39
Metcalfe, Robert, 91. *See also* Metcalfe's law
Metcalfe's law, 87, 91–93, 99, 194
Metrics
  defined, 12, 24 (*See also* Metrics, techonomic)
  techonomic, 27, 45–84
    agriculture, 73–74
      productivity gains, 74
    annual per capita energy consumption, 55
    communication, 64–70, 81
    community, 70–79, 81
    community health, 74–77
    computation, 57–63, 81
    constant techonomic metric measure, cost vs. performance, 47
    cultural approaches to endeavor, 45
    defined, 24–27, 45
    digital photography techonomic metric, 48–51
      camera development, 50
      camera sales, 50
    endeavor trends, 45
    energy, 54–57, 81
    environment, 74–77
    healthcare expenditures, 76
    interdependence, 79–81
    media, 77–79
      influence of, 77–79
    metric, techonomic, 81
    military, influence on technology advances, 51–54
    spiritual condition, 79
    sustainability, techonomics of, 71–73
    technical approaches to endeavor, 45
    world population, 71–72
Mexico, student performance in, 169
Micro-multinational virtual company, 194

Microsoft, 119
  decoupling of key products, 149
Military, influence on technology advances, 13–15, 51–54
Molecular embossing, 188
Moore, Gordon, 88
Moore's Law, 87–91, 98, 175–176, 189, 194
Motors, superconducting, 190

N

Naisbitt, John, 133
Nanocircuitry, 188–189
Nanogels, 188
Nanotechnology, 187–189
Nanotubes, 188
National survival, military promotion of technology, 14
"The Nature of Firm," 19
Near-eye display systems, 185–186
Netherlands, student performance in, 169
New Zealand, student performance in, 169
Newton, Sir Isaac, 11, 31
Norway, student performance in, 169
Nuclear energy sources, 137–139

O

Oak Ridge National Laboratory, particle accelerator, 188
OECD-PISA. *See* Organization for Economic Cooperation and Development-Program for International Student-Assessment
Organic evolution
  Darwin's theory of, 6–10
  organizational evolution, 16
Organization, defined, 12
Organization for Economic Cooperation and Development-Program for International Student-Assessment, 168–171
Organizational evolution
  four-square principal for, 38
  human
    life cycles of, 7
    technological advancement, 31–44
      communication, evolution in, 39–40
      community, evolution in, 40–41
      computation, evolution in, 39
      energy, evolution in, 37–38
      four-square principle, 36–41
        for individuals, 41

Index

    for organizations, 42
    history of technological breakthroughs, from prehistory to 1500 AD, 33
    human evolution, four-square principle for, 37
    mental evolution, 39
    organizational evolution
        four-square principal for, 38
        timeline, 32–36, 41
    physical evolution, 37–38
    social evolution, 39–40
    spiritual evolution, 40–41
    technology timeline, 41
  from technological advancement, timeline, 32–36, 41
  techonomics, 6–10, 16
  timeline, 32–36, 41
Organs, replacement, 192
Outsourcing, 95
Overview of techonomics, 3–18

P

Parallels in evolutionary frameworks, 9, 16
Particle accelerator, 188
Pate, Thomas, 167
"Perfect information," importance of, 22–24
Personal telemedicine, 192
Pharmaceuticals, personalized, 147
Physical evolution, 37–38
Planting, vertical, care methods, 185
Plants, genetically engineered, 185
Plasma, lighting, 189
Poland, student performance in, 169
Population data, 71–72
Portugal, student performance in, 169
Postindustrial challenges, 155–204
  future of techonomics, 183–204
  techonomic market crises, 157–182
Predictive medicine, 191
Pride of Southland Band, University of Tennessee, 167
Printing, three-dimensional, 187
Protectionism, economic, 148–150
Punctuality, hidden transaction cost, 21
Purpose of techonomics, 3–18

Q

Quality, hidden transaction cost, 21

R

Renewable energy resources, 133–140, 151–152
  biological sources, 134
  chemical sources, 136–137
  contemporary energy technologies, 138–140
  cyclical sources, 134–136
  hybrid vehicles, 139
  nuclear power, 138–139
  nuclear sources, 137–138
  solar-voltaic electricity, 139
  solar voltaic sources, 135–136
  wave sources, 135
  wind sources, 135
Replacement organs, 192
Risk, hidden transaction cost, 22
Roosevelt, Theodore, 157
Russia, student performance in, 169

S

Self-interest, as fundamental assumptions in techonomics theory, 13
Services, virtual, 195
Sheathing, plasma, 189
Slovak Republic, student performance in, 169
Smith, Adam, 196–197, 205
Social evolution, 39–40
Solar spectrum lighting, 184
Solar voltaic energy sources, 135–136, 139
Spain, student performance in, 169
Spallation Neutron Source, 188
Specialization, 147–152
  emerging techonomic trends, 152
Speed business, 119
Spiritual evolution, 40–41
Stereo projection, 185
Sterilization, plasma, 189
Superconductors, 190–191
Sustainability, techonomics of, 71–73
Sweden, student performance in, 169
Switching, hidden transaction cost, 22
Switzerland, student performance in, 169

T

Technical approaches to endeavor, 45
Technological advancement, organizational/human evolution, 31–44
  four-square principle, 36–41
    for individuals, 41
    for organizations, 42

history of technological breakthroughs, 33
human evolution, four-square principle for, 37
mental evolution, 39
organizational evolution
    four-square principal for, 38
    timeline, 32–36, 41
physical evolution, 37–38
social evolution, 39–40
spiritual evolution, 40–41
technology timeline, 41
Technology, derivation of term, 11
Technology timeline, 41
Techonomic metrics. *See* Metrics, techonomic
Telemedicine, personal, 192
Temperature, extremes of, 189–191
Theory of Organic Evolution, 1–10, 16, 197, 206
Three-dimensional entertainment, 185–187
Three-dimensional printing, 187
Timeliness, importance of, 23
Transactions, 19–30
    accuracy, 22
    availability, 21
    Coase, Ronald, 19–20
    completeness, 23
    contributors to costs, 26
    cost, 23
    cost analysis, 19–21, 26
    defined, 12
    hidden transaction costs, 21–22
    inventory, 21
    metric, defined, 24
    metrics, techonomic, defining, 24–27
    "The Nature of Firm," 19
    "perfect information," importance of, 22–24
    punctuality, 21
    quality, 21
    risk, 22
    switching, 22
    techonomic metric, 27
    timeliness, 23
    transport, 21

Transport, hidden transaction cost, 21
Turkey, student performance in, 169
Twenty-first century, techonomics at turn of, 85–154. *See also* Postindustrial challenges
    emerging business models, 103–132
    laws of techonomics, 87–102
    trends in techonomics, 133–156

U

Ubiquitous computing, law of. *See* Moore's law
Ubiquitous global network, law of. *See* Metcalfe's law
Ubiquitous sensory network, 146–147
USA, student performance in, 169

V

Vehicular media expansion, 146
Vertical planting, care methods, 185
Video displays, 185
Virtual companies, 194–195
Virtual media, 125–127
Virtual reselling, 123–125
Virtual retail, 120–123
Virtual services, 195
Virtual world, 192–196
Virtualization, contemporary examples, 193

W

Wal-Mart, as business model, 109–114
Walgreens, as business model, 114–116
Wave energy sources, 135
Wind energy sources, 135
World population, 71–72